SCHAUM'S OUTLINE OF

THEORY AND PROBLEMS

OF

UNDERSTANDING

CALCULUS CONCEPTS

ELI PASSOW, Ph.D.

Professor of Mathematics
Temple University
Philadelphia, Pennsylvania

SCHAUM'S OUTLINE SERIES

McGRAW-HILL

New York San Francisco Washington, D.C. Auckland Bogotá Caracas Lisbon
London Madrid Mexico City Milan Montreal New Dehli
San Juan Singapore Sydney Tokyo Toronto

Eli Passow is a Professor of Mathematics at Temple University, having received a B.S. in Mathematics from the Massachusetts Institute of Technology, and M.A. and Ph.D. degrees from Yeshiva University. He has taught at Yeshiva University, Bar-Ilan University, and The Technion, and has published over 35 papers in Approximation and Interpolation Theory. He is a member of the Mathematical Association of America.

Schaum's Outline of Theory and Problems of
UNDERSTANDING CALCULUS CONCEPTS

4 5 6 7 8 9 10 11 12 13 14 15 16 17 18 19 20   PRS/PRS   9 0 1 0 9

ISBN 0-07-048738-3

Sponsoring Editor: Arthur Biderman
Production Supervisor: Donald F. Schmidt
Editing Supervisor: Maureen Walker

Library of Congress Cataloging-in-Publication Data

Passow, Eli.
    Schaum's outline of theory and problems of understanding calculus
  concepts / Eli Passow.
        p.   cm.   — (Schaum's outline series)
    Includes index.
    ISBN 0–07–048738–3
    1. Calculus—Outlines, syllabi, etc. 2. Calculus—Problems,
  exercises, etc.   I. Title.
  QA303.P284   1996
  515′.076—dc20   95–47937
                                                      CIP

McGraw-Hill

A Division of The McGraw·Hill Companies

Dedicated to

Chaya

Shuli, Nati and Dani

and

In Memory of

my parents

Meyer and Clara Passow

Acknowledgements

I owe a great deal to two very special people. Rachel Ebner did a magnificent job of editing my manuscript, which substantially enhanced its clarity. Lev Brutman taught me the intricacies of T_EX, and provided important advice and encouragement throughout the writing of this book. Thanks also to my Sponsoring Editor, Arthur Biderman, for his patience and help, the staff at Schaum's, and to the reviewers, for their many useful suggestions.

Preface

Dear Student,

Feeling butterflies at the thought of beginning your Calculus course? Or worse? Well you're not alone. There's a long tradition of students suffering from 'Calculus Syndrome.' I myself took Calculus in 1968 and hated every minute of it. I somehow survived the agony, passed the course barely respectably, and sold that fat, heavy textbook right away — who needed memories of humiliation? I went on to major in languages so that I could never again be traumatized by numbers.

And so, when 25 years later Dr. Eli Passow approached me to edit a companion volume to Calculus texts, I laughed and told him straightforwardly: "You've come to the wrong address. I'm allergic to mathematics."

But Dr. Passow convinced me that I was just the person he was looking for. He had developed a simple and clear conceptual approach to Calculus and wanted to test out his ideas on someone who was absolutely certain that Calculus could never penetrate her brain. Okay, so he had the right address after all.

I approached "Understanding Calculus Concepts" with all the skepticism Dr. Passow could have wished for — it came quite naturally — but I was quickly surprised. Lightbulbs of understanding started going on in my head that should have lit up a quarter century ago! Dr. Passow's $\mathcal{A}$—$\mathcal{R}$—$\mathcal{L}$ approach (Approximation, Refinement, Limit — I'll let him explain it) to all the major concepts of Calculus quickly led me to the crucial awareness of the unity of the subject, a soothing substitution for my past experience of Calculus as a zillion different kinds of problems, each demanding a different technique for solution. With the $\mathcal{A}$—$\mathcal{R}$—$\mathcal{L}$ method, the approach to a wide variety of problems is the same; learning takes place in a comprehensive way instead of in unrelated fragments.

And most important, in Dr. Passow's writing I heard the voice of a real teacher, the teacher we all *wish* we had, especially for Calculus. He has a gift for simplifying and a knack for presenting illuminating illustrations and examples that help each concept click into place, even in a mind that was determined not to understand.

So, don't panic. You're in good hands. 'Calculus Syndrome' has a cure and you're holding it right now.

Good luck!

Rachel Ebner

Contents

Chapter 1

What It's All About

1.1 Introduction

For many students, calculus is a frightening subject, frightening even before the course begins! For one thing, you are handed (or, more correctly, you buy at a handsome price) an intimidating 1000-page volume chock-full of complicated material for which you will be responsible over the next two or three semesters. The material itself has the reputation of being far more difficult than any mathematics you've previously studied, filled with a bewildering assortment of apparently unrelated techniques and methods. Last but not least, the high failure rate in calculus is notorious; students repeating the course are clearly in evidence around you. So you're probably asking yourself, "What am I doing here?"

Yet, as we'll see in section 1.3, calculus need not be a difficult course, at least not when properly presented. Nonetheless, you might ask yourself, "Why should *I* bother studying this stuff, anyway?" I know that it's a requirement for many of you, but I'll try to come up with a better answer than that.

What would a world without calculus look like? Well, on the positive side, you wouldn't be taking this course. But you would be living in a world without most of the modern inventions that (for better or for worse) we rely upon: Cars, planes, television, VCRs, space shuttles, nuclear weapons, and so forth. It would also be a world in which much of medicine would be practiced on the level of the 17th Century: No X-rays, no CAT scans, nothing that depends upon electricity. And it would be a world without the statistical tools which both inform us and allow us to make intelligent decisions, and without most of the engineering feats and scientific and technological advances of the past three centuries. For some strange reason that we do not fully understand,

nature obeys mathematics.

And among all of mathematics, *calculus* stands out for its applicability, its relevance to the practical world. So even if you're not entering a technical or scientific field, calculus is (or should be) a part of your *general* education, to enable you to understand the above features of the modern world, just as everyone — including mathematicians and scientists — should be exposed to Shakespeare, Plato, Mozart, and Rembrandt to appreciate the richness of life. However, calculus is such a vast subject and so filled with technical mathematical details, that it will take some time before you can understand how many of these applications of calculus arise. Be patient; we'll get there.

Among other things, calculus includes the study of motion and, in particular, *change*. In life, very few things are static. Rivers flow, air moves, populations grow, chemicals react, birds fly, temperature changes, and so on and so on. In fact, in many situations, the *change* of a quantity is at least as important as the quantity itself. For example, while the actual temperature of a sick person is important, the *rise* or *fall* (change) in temperature may be a crucial indicator of whether an illness is worsening or coming under control. In another context, companies ABC and XYZ may currently have identical profits. However, if ABC's profits are increasing *rapidly*, while those of XYZ are *barely* increasing, then this difference in the *rates of change* of the profits of the two companies will make the Board of Directors of ABC much happier than those of company XYZ.

In fact, you are probably already familiar with some aspects of calculus, since notions of change have entered our everyday vocabulary, as well as that of many disciplines in the sciences, social sciences, and business. Terms such as speed and acceleration from physics, inflation rate and marginal profit from economics, reaction rates from chemistry, and population growth rate from biology all involve change and, as we'll see, they are all examples of one of the most important concepts in calculus, the *derivative*, which we will study in Chapter 2.

1.2   Functions

Change involves a relationship between two quantities that vary. For example, if you deposit $100 in a bank, then the *amount* you will have in your account a year later depends upon the *interest rate* the bank pays. If the rate is fixed, but your deposit differs from $100, then the *amount* depends upon the size of your initial *deposit*. A $200 deposit will grow to an amount twice that of a $100 deposit, while a $1000 deposit will be 10 times as large.

Similarly, if you drop a ball from a window, then its *speed* when it hits the ground depends upon the *height* of the window above the ground. So does the *time* it takes for the ball to reach the ground.

Each of these examples involves two quantities that can vary and, as one of them changes, the second one usually does as well. We call such quantities *variables*, and the *relationship* between the variables is called a *function*. Referring back to our examples, we say that the amount of money in your bank account is a *function* of the interest rate if the size of the deposit is fixed, but is a *function* of the deposit if the interest rate is fixed. The speed of the ball when it hits the ground is a *function* of the height from which it was dropped, and so is the time it takes to reach the ground.

Now suppose that one variable is a function of a second one. As is usual in mathematics, we find it convenient to introduce symbolic notation to represent this relationship. The usual way of denoting a variable is by a letter, if possible, one which relates to the specific variable. In the bank account example we might let A stand for the amount in the account and i for the interest rate. We would then say that "A is a function of i." But having to write "A is a function of i" is still too clumsy, and we want an even shorter notation. This is done symbolically by writing $A = f(i)$ (read "A equals f of i"), with the letter f being the *name* of the function, and $f(i)$ its *value* at i.

For example, if the interest rate, i, is 6%(.06) and \$100 is deposited, then the amount A in the account at the end of one year will be \$106. Thus, we write $f(.06) = \$106$. Similarly, $f(.08) = \$108$, since the \$100 deposit will grow to this amount if the interest rate is 8% (.08).

When studying functions in a general context, where the variables may not have specific meaning, we often use the familiar x and y, writing $y = f(x)$, as in $y = x^2$ or $y = \sin x$.

Important: A function is allowed only one value for any x.

Keep in mind, also, that not every function can be expressed by a neat formula; sometimes the relationship consists of measurements taken from time to time. An example is the measure of inflation known as the Consumer Price Index (CPI), which is computed monthly by the Department of Labor. No formula can be used to describe the CPI or to predict its value precisely in subsequent months. Nevertheless, the CPI can be analyzed using techniques that we'll develop in this course.

The equation $y = f(x)$ is an *algebraic* entity. However, if we plot in the plane the set of all points (x, y) whose coordinates satisfy this equation, then the curve we obtain is called the *graph* of the equation $y = f(x)$ (or the graph of the function, f). Plotting the graph allows us to apply geometrical ideas in the study of functions. For example, the graph of every equation of the form, $y = mx + b$ is a straight line. (As a result, such equations are called linear equations.) Frequently, the geometrical picture gives us insight lacking in the algebraic formula.

Now, the concepts of calculus apply to functions in general. Here's an example (you haven't been introduced to the terminology yet, but don't worry about that): If two functions are differentiable, then their sum is also differentiable. In symbolic form, we write: If f and g are differentiable, then $f + g$ is differentiable. This is a general statement which is valid for *any* two functions, f and g. Calculus, whose purpose is to analyze functions in many different ways, is filled with statements of this type.

But calculus also goes beyond the general to the particular. There are a number of functions which appear so frequently in important applications that they deserve 'names,' rather than just the anonymous f or g. Examples include polynomials (such as x^2 or $3x^5 + 4x^2 - 7x + 2$), trigonometric functions ($\sin x$ or $\cos x$), and exponential functions (2^x or 10^x). The functions which are prominent enough to warrant special names are called the elementary functions; a list of the ones you will encounter is contained in the table 'The Cast of Characters,' which follows. Although much of your work in calculus will involve learning how to manipulate the elementary functions, the *general results* that underlie these manipulations constitute the *heart* of the course. Learning to do computations with the elementary functions is part of your job; equally important is understanding the broad general concepts.

The Cast of Characters	
Function Type	**Examples**
Linear	$2x + 5, \quad -3x + 4$
Power	$x^4, \quad 6x^3$
Polynomial	$x^3 - 2x + 1, \quad 3x^5 + 2x^4 - 7x^3 + x + 4$
Rational	$(x^2 + 3x - 2)/6x^3 - 1)$
Algebraic	$\sqrt{x}, \quad (x^2 + 5x)^{1/3}$
Trigonometric	$\sin x, \quad \cos x, \quad \tan x$
Inverse trigonometric	$\sin^{-1} x, \quad \tan^{-1} x$
Exponential	$10^x, \quad e^x, \quad 2^{-x}$
Logarithmic	$\log_{10} x, \quad \ln x$
Hyperbolic	$\sinh x, \quad \cosh x$
Inverse hyperbolic	$\sinh^{-1} x, \quad \tanh^{-1} x$

1.3   Calculus: One Basic Idea

I said earlier that calculus need not be a difficult course. The reason for this is that calculus is founded upon just one fundamental and easily understood idea, which threads its way through almost every topic we will encounter. The idea is that of *approximation*. Many of the concepts in this course evolve in the following pattern: We begin with a familiar idea, which works well in relatively *simple* cases. We wish to generalize this idea to a more complicated situation. However, we do not have the mathematical tools to tackle the more difficult problem, so rather than attempting to solve it *exactly*, we choose, instead, to be temporarily satisfied with an *approximate* solution. This approximation is then *refined*, or improved, so as to provide a better estimate of the desired quantity. We continue to refine the approximation, finally reaching a concept known as the *limit*. All of these ideas will be made precise when we get to specific examples, but for now, take comfort from the fact that much of calculus can be broken down into a simple three-stage process:

Approximation — Refinement — Limit

which we abbreviate as $\mathcal{A}$—$\mathcal{R}$—$\mathcal{L}$. The details, of course, differ from case to case, but the basic pattern repeats throughout calculus. As you begin to understand and work with $\mathcal{A}$—$\mathcal{R}$—$\mathcal{L}$, you will experience calculus as a unified subject, rather than just as a collection of techniques for solving a variety of problems.

Now, the idea of approximation is very common in everyday life. We use it when trying to find an entry in a dictionary, when checking the temperature of the water in a bathtub, or when weighing ourselves on a doctor's scale. In each case, our initial approximation is *refined* several times, each time moving closer (hopefully!) to our goal, until we reach a satisfactory conclusion—the correct word, the right temperature, the correct weight. Scientific theories are also approximations which describe physical situations with a certain degree of accuracy, and which are continually being refined to more exactness. So the idea of approximation is not new to you.

1.4 Conceptual Development

The development of any concept, not just in mathematics, goes through a number of stages. For mathematical concepts these stages include:

- Motivation

- Definition

- Notation

- Computational Techniques

- Applications

Since these distinctions may be unfamiliar, we'll elaborate a bit. But first, we need to recognize that a mathematical concept is *much more* than just a procedure or operation for solving a particular problem. Generally, the concepts of calculus are deep ideas, which took the mathematical community many *centuries* to discover, develop, and understand. These ideas have widespread applications; new ones continue to arise, more than 300 years after the foundations of calculus were discovered by Newton and Leibniz. Let's now walk through the five stages mentioned above.

Motivation: What need existed which led to the creation of this concept? Often, several apparently *different* problems, drawn from a *variety* of fields within and outside mathematics, turn out to be closely related, and lead us to the formulation of a *general* concept. Similar situations occur outside of mathematics. For example, consider the notion of *democracy*. If there were but *one* democratic country in the world (all the others being monarchies or dictatorships), it is doubtful if the word 'democracy' would even exist. However, there are *many* democratic countries, although their forms of government differ significantly (parliamentary systems, presidential systems, and so forth). As a result, the *general* concept of democracy is studied intensively.

Definition: After recognizing the similarities in the problems that motivated the study, we eventually extract the *concept* or *idea* common to all of them, and give it a *name*. That's all a definition is. The definition is general, including as special cases those problems which originally motivated the development of the concept.

Notation: Symbolic notation plays a powerful role in conceptual development. Notation certainly provides a concise form of shorthand, but, more importantly, good notation enhances our understanding of a concept and facilitates our computational abilities. For example, compare the multiplication of 387 by 834 in our system of numeration with the amount of work the Romans would have to do in *their* system (CCCLXXXVII times DCCCXXXIV).

Computational Techniques: As we will see over and over, the formal definitions are generally too clumsy to be of much use when we actually try to apply them to specific cases. As a result, we search for alternate procedures or *shortcuts*, which make the computations more efficient. (**Aside:** This problem occurs outside of mathematics, as well. For example, here is a dictionary definition of the word 'cat': "A long-domesticated carnivorous mammal that is usually regarded as a distinct species though probably ultimately derived by selection from among the hybrid progeny of several small Old World wildcats, that occurs in several varieties distinguished chiefly by length of coat, body form, and presence or absence of tail, and that makes a pet valuable in controlling rodents and other small vermin but tends to revert to a feral state if not housed and cared for." Now, how much use will this definition be to us if we see an animal on the street and ask whether it actually *is a cat*?)

Applications: Most of the concepts in calculus have applications that go far beyond the problems that provided the original motivation. Many of these applications arose decades or centuries after the concept was uncovered. For example, the concept of the integral, which we'll study in Chapter 4, was motivated by the need to solve specific problems in physics and astronomy, dealing with planetary motion. Soon after that, it was found to have other applications, among them computing the volumes of certain solids. But in the 1960's, 300 years after the discovery of calculus, an application of the integral, known as the Fast Fourier Transform (FFT), was developed. The FFT has important consequences in biomedical engineering, the design of aerodynamically efficient aircraft, and many others. The original paper which introduced the FFT has been cited in well over 1000 articles in a large variety of scientific journals. We will have more to say about the FFT in Chapter 5.

> **A true knowledge of calculus is possible only if you learn to distinguish these various aspects of the concepts of calculus. This book will assist you in reaching that goal.**

1.5   About this Book

The purpose of this book is to aid you in learning the *concepts* of calculus. It is intended as a *companion* to your text. I've written it in a casual style, with full explanations, and many examples and diagrams to illustrate and help you visualize the concepts. In it, you will find relatively few details or techniques, since they can be found in the text you are using. Many standard calculus books do an adequate job of presenting this material; I hope that yours is one of them! But to *understand the concepts*, you will find it useful to return to this book on many occasions. I suggest that you read each chapter just *before* you begin the parallel one in the text. Then review before a test to help fix the concepts in your mind. Finally, go through the relevant chapters of this book one more time at the end of the semester in preparation for the final exam. In particular, I suggest that you make frequent use of the table which is found at the end of this chapter. It reveals many of the topics you will be covering in the course, and shows how the 'difficult' concept is obtained from the 'easy' or 'known' one by approximation. It thus serves as both an overview of what will be coming, as well as a summary of the material.

At various places in the book you will notice a box in the margin of the page, just like the one here. The purpose of these boxes is to help you coordinate the material in *this* book with that in *your* text. The boxes occur in places where I refer you to your text for the proof of a theorem, computational procedures, additional examples, or the development of a topic that will not be covered in this book. You will find it useful to fill in the box with the page number of the corresponding section in your text, which contains details and extensions of the concepts introduced here. If you do this consistently, then you will find it quite easy to jump back and forth between the books, which will be of great help, *especially when you are preparing for exams.* In other places, you will find the letters $\mathcal{A}, \mathcal{R}$, or $\mathcal{L}$ in the margin. As you might expect, these symbols alert you to the fact that the $\mathcal{A}$—$\mathcal{R}$—$\mathcal{L}$ process is underway, and take you through these three steps.

The book is structured as follows: Each of the three main chapters, Chapter 2 (The Derivative), Chapter 4 (The Integral) and Chapter 7 (Infinite Series) is divided into 5 sections, following the format: Motivation, Definition, Notation, Computational Techniques, and Applications. However, while we do mention the numerous applications of the derivative and integral in Chapters 2 and 4, in each case we devote a separate chapter to the development of the details of *two* of the applications, which are found in Chapters 3 and 5, respectively. Chapter 6 includes important additional topics which involve the integral. You can use this book no matter what order your instructor arranges the material. It is also worth noting that *all* of the topics in this book, even the applications, adhere to the Approximation—Refinement—Limit framework ($\mathcal{A}$—$\mathcal{R}$—$\mathcal{L}$).

A word about the problems. There are both solved problems and supplementary problems in each chapter. Some are computational, others are of a conceptual nature, and still others are extensions of theoretical material not contained in the body of the text.

In conclusion, constantly keep in mind the organizing framework of both the book and calculus: First, the breakdown of each concept into its five stages,

- Motivation

- Definition

- Notation

- Computational Techniques

- Applications

and second, of course,

- Approximation

- Refinement

- Limit

$$\mathcal{A}\text{---}\mathcal{R}\text{---}\mathcal{L}$$

Known Concept	Undefined Concept	Approximation Technique
Secant line	Tangent to a curve	Secants approximate tangent
Slope of a line	Slope of a curve	Slopes of secants approximate slope of tangent
Average velocity	Instantaneous velocity	Average velocities approximate instantaneous velocity
Area of rectangle	Area of curved region	Rectangles approximate region
Length of a line	Length of a curve	Broken lines approximate curve
Work (constant force)	Work (variable force)	Constant forces approximate variable force
Volume of cylinder	Volume of solid	Cylinders approximate solid
Integral (proper)	Improper integral	Proper integrals approximate improper integral
Addition	Infinite series	Finite sums approximate infinite series
Root of linear function	Root of $f(x)$	Newton's method
$\int_a^b (cx + d)\, dx$	$\int_a^b f(x)\, dx$	Trapezoidal Rule
$\int_a^b (cx^2 + dx + e)\, dx$	$\int_a^b f(x)\, dx$	Simpson's Rule

Chapter 2

The Derivative

2.1 Motivation

Let us now put our organizing principles to work. We consider two apparently unrelated problems which will motivate our first concept, known as the derivative. The first is geometric, dealing with the slope of curves, while the second, instantaneous velocity, is physical.

2.1.1 Slopes and Tangent Lines

> **What we know:** Slope of a straight line.
>
> **What we want to know:** Slope of an arbitrary curve.
>
> **How we do it:** Approximate the curve with certain straight lines.

We are familiar with the notion of the slope of a straight line. Every line has associated with it a single number which represents the slope. Intuitively, the slope represents the steepness of the line; a line with large positive slope is steeper than one with smaller positive slope, while a line with negative slope is falling as we move from left to right (Figure 2-1). Now why should anyone other than a mathematician

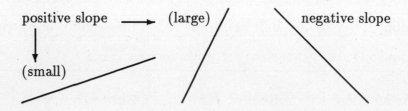

Figure 2-1: Slopes of various lines

be interested in slopes? Well, the slope of a line involves *change* and, as we saw in Chapter 1, change plays a prominent role in calculus. But just how are slope and change connected? To answer this question, let's recall the definition of the slope of a straight line. If a line passes through the points (x_0, y_0) and (x_1, y_1) (Figure 2-2), then its slope is defined by

$$m = \frac{y_1 - y_0}{x_1 - x_0}. \tag{2.1}$$

(Nobody seems to know exactly why we use the letter m for the slope of a straight line rather than S, but this notation is common and we generally adhere to it.)

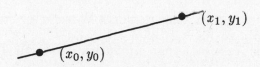

Figure 2-2: Slope of a line

Let's look at the slope m that we've just calculated. The numerator, $y_1 - y_0$, is the *change* in y which occurs when x changes from x_0 to x_1. Mathematicians often use the symbol Δ to denote change. Thus, we write $y_1 - y_0 = \Delta y$ (read "delta-y") and $x_1 - x_0 = \Delta x$. Using this notation, we obtain from (2.1)

$$m = \frac{y_1 - y_0}{x_1 - x_0} = \frac{\Delta y}{\Delta x}.$$

Now $\Delta y/\Delta x$ tells us how fast y is changing *with respect to* x. In other words, it represents the *rate of change* of y with respect to x. For example, if $\Delta y/\Delta x = 2$, then y is increasing twice as fast as x, while if $\Delta y/\Delta x = -3$, then y is *decreasing* by three units as x increases by one unit. In particular, if, say, the line is the graph of the profits of a company over a period of several years, then the slope represents the *change* in profits, which may be increasing or decreasing, rapidly or slowly, depending on the *sign* and *size* of the slope. The notion of rate of change is fundamental and widespread throughout the sciences, engineering, business, and the social sciences, and includes such topics as velocity and acceleration in physics, changes in profit and the inflation rate in business and economics, and reaction rates in chemistry. A discussion of the many applications of rates of change in various fields will be found later on in this chapter (page 35).

However, real-world situations rarely generate straight line graphs. Few companies have profits that regularly increase or decrease linearly. So we are faced with the problem of extending the notion of slope to general *curves*. To do so, let us recall that for any specific line, the slope is independent of the choice of points; that is, the slope is the same everywhere on the line. Thus, the slope of a straight line is a single number, which can be calculated by choosing *any* two points on the line, (x_0, y_0) and (x_1, y_1), and applying (2.1). A glance at a picture, however, makes it clear that we cannot expect a single number to represent the steepness of a curve, which changes from point to point. For example, it is obvious that the curve in Figure 2-3 is much steeper near

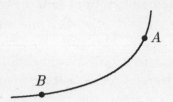

Figure 2-3: The steepness of a curve varies

point A than near point B. So does it make sense to speak about the slope of a curve *at a point*? In other words, can we assign *numbers* to the curve at the points A and B which reflect the fact that the curve is steeper at A than it is at B? We now come to one of the main features of calculus.

> When faced with a difficult problem, we initially abandon our attempt to solve it *exactly* and, instead, look for an *approximate* solution.

How can this principle be applied in our situation? Well, currently all we know about slopes is how they work for *straight lines*, so this information must somehow be used in the solution of the problem. In the following, we shall see that this knowledge is sufficient to accomplish our purpose.

We wish to define the slope of any curve at a point. This is expressed mathematically as follows: Suppose that the curve can be represented by the equation $y = f(x)$, and the point, P, by the coordinates (x_0, y_0), where x_0 is any value of x and $y_0 = f(x_0)$ (Figure 2-4). We will define the slope of the curve at P to be equal to the slope of the

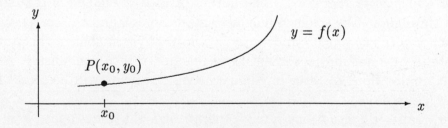

Figure 2-4: Graph of the function $y = f(x)$

line *tangent* to the curve at P. But what do we mean by the tangent line? This term is familiar from high school geometry, where the tangent to a *circle* is defined as a line which touches the circle at exactly one point.

This definition of the tangent, while adequate for the circle, does not extend directly to other curves, and it is another of its properties which is more useful in calculus: The tangent to the circle at P is the line which 'stays closest' to the circle near P, or, in other words, among all lines passing through P, it is the one which *best approximates* the circle near P. In order to understand just what we mean by this, we introduce another line which will be of importance to us, the *secant*, a line which touches the circle at *two* points (Figure 2-5). Now, let's examine a small section of the circle in Figure 2-5 near

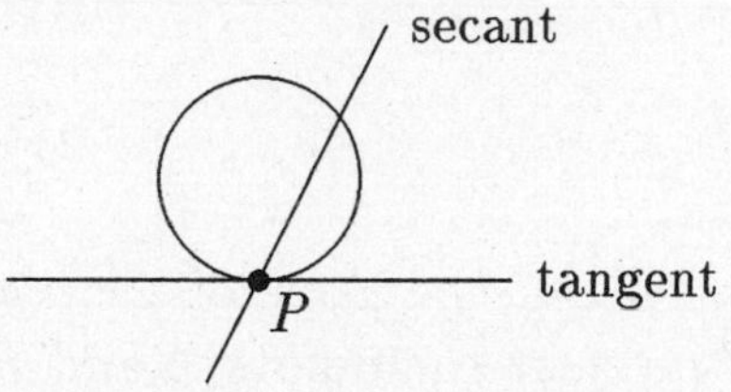

Figure 2-5: Tangent and secant to a circle

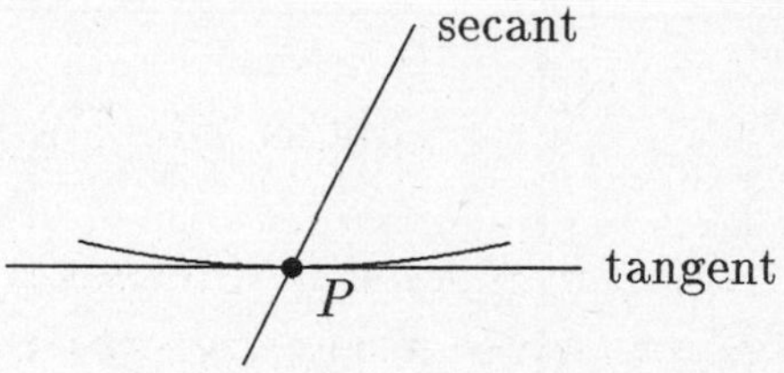

Figure 2-6: Magnified view of an arc of the circle

P with a very powerful magnifying glass. What we see resembles Figure 2-6. In words, while the tangent and arc of the circle are almost indistinguishable near P, the secant may be clearly differentiated from the circle. It is in this sense that the *tangent* 'best approximates' the circle near P.

The Greeks gave a definition which is similar to this: The tangent, T, is a line with the property that it is impossible to 'fit' another line through P lying between T and the curve (Figure 2-7). In other words, the tangent is a line which 'hugs' the curve.

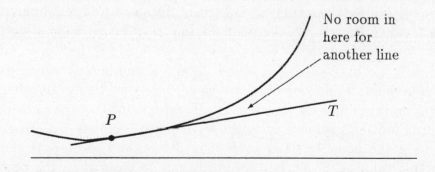

Figure 2-7: Greek definition of a tangent

Here's another approach to the tangent. Suppose a car is moving along a curved road at night. In what direction do the headlights point? That's right, in the direction of the tangent to the curve.

So the tangent line determines the direction *of the curve.*

We now turn to the problem of *computing* the slope. We will assume, initially, that for our curve the tangent line exists (it doesn't always), and our question is merely that of *finding* its slope, which we denote by m. We use the following procedure (Figure 2-8). Choose a second point on the curve reasonably close to P, say $Q = (x, y)$. Analogous

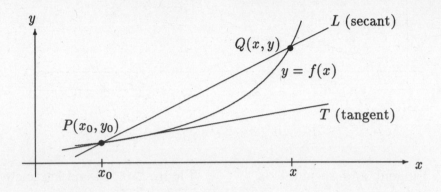

Figure 2-8: Initial approximation of the slope of T

to the circle, we call the line through the two points P and Q a *secant*, which we denote by L. Because we know two points on L, we can find its slope, S, which is given by

$$S = \frac{y - y_0}{x - x_0}. \tag{2.2}$$

(Why do we denote the slope of the secant line by the letter S, rather than by m, which is the customary symbol for the slope of a line? We do this because we will be discussing the slopes of *both* the tangent and secant lines and we need different symbols to distinguish between them.) We consider the line L to be an *approximation* to the tangent line, T, and its slope, S, to be an *approximation* to m, the slope of T.

Where are we? We have produced a line, L, and a number, S, which presumably are close to the tangent, T, and its unknown slope, m, respectively. In other words, L and S are *initial* approximations to T and m. However, since we may not be satisfied with these approximations, we now enter the second stage, *refinement*. To get a better estimate, we *move* the point Q closer to P than before, and recompute the slope, S. We hope that this value of S is a better approximation to m than the previous one (Figure 2-9). The refinement process now continues by moving the point Q even closer along the curve toward P (Figure 2-10). We expect that the corresponding secants and slopes obtained are successively better approximations to T and m (Figure 2-11). Indeed, by choosing Q sufficiently close to P, we hope that the approximation will become as accurate *as we wish*. But how do we finally obtain the tangent, T, and the *exact* value of the slope? For this, we enter the third stage, called *passing to the limit*.

We will not discuss any of the technical difficulties which can crop up in this stage, because our purpose is to achieve a broad understanding of the topic, rather than to carry out an exhaustive logical analysis. More details can be found in your text.

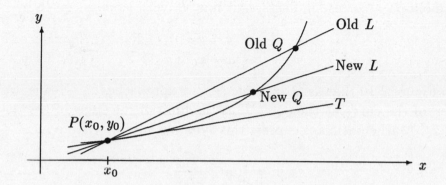

Figure 2-9: Refined approximation of the slope of T

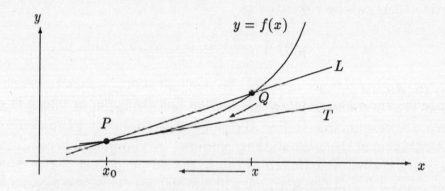

Figure 2-10: Further refinement of the approximation

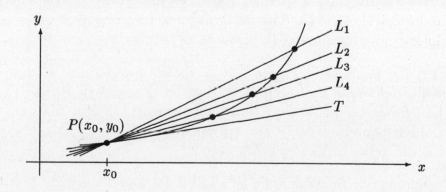

Figure 2-11: The secants are approaching the tangent

We adopt, instead, an intuitive approach. Just imagine that the process of moving Q along the curve closer and closer to P continues *indefinitely*. For each choice of the point, Q, we obtain a corresponding secant line, L, with slope S, given by

$$S = \frac{y - y_0}{x - x_0}.$$

(2.3)

Notice from Figure 2-10 that as Q approaches P, x approaches x_0. Hence, we can define the slope, m, of the curve at the point P as the limit of the slope of the secant L as x approaches x_0. Mathematicians express this symbolically by writing

$$m = \lim_{x \to x_0} \frac{y - y_0}{x - x_0},$$

(2.4)

which is read as follows: *m equals the limit, as x approaches x_0, of the fraction $y - y_0$ divided by $x - x_0$.*

Now recall that the equation of the curve is $y = f(x)$, so that $y_0 = f(x_0)$. As a result, equation (2.4) can be rewritten as

$$m = \lim_{x \to x_0} \frac{f(x) - f(x_0)}{x - x_0},$$

(2.5)

provided this limit exists.

We'll soon give a concrete example illustrating these general ideas, but first let's review. Since we were unable to find the slope of the tangent to the curve at P, we decided to seek an *approximate* solution, obtained by choosing a second point, Q, on the curve. Having two points allowed us to use the simple notion of the slope of a line, since the two points, P and Q determine a line, which we called the secant, L. We took the slope, S, of the secant to be an *initial approximation* to the quantity, m, we were seeking, and then *refined* our approximation by choosing points Q closer and closer to P. Finally, we obtained the exact value of the slope by passing to the *limit*.

Observe how we have gone through the three important stages, *approximation*, *refinement* and *limit* ($\mathcal{A}$—$\mathcal{R}$—$\mathcal{L}$). This pattern appears repeatedly in calculus, although the details differ (sometimes markedly) from topic to topic.

Remark 2.1 The point Q in Figure 2-10 is shown approaching P from the right, which is called a *right-hand limit*. Similarly, if we let Q approach P from the *left*, then we obtain the *left-hand limit*. We say that *the limit* exists at P if and only if *both* the right-hand and left-hand limits exist and are equal to each other. This requirement will make a crucial difference in Example 2.3, page 20.

Before beginning our first example, let's talk a bit about mirrors. It is a principle of physics that if a ray of light hits a *straight* mirror at an angle, a, then it reflects off the mirror at the same angle: The angle of reflection equals the angle of incidence (Figure 2-12). Now what happens if the mirror is not a straight line, but a curve. Then the same principle holds: The angle of reflection equals the angle of incidence. But, hold on! Angles have *two* sides, both of which are straight lines. So what do we mean by the angle between a line and a *curve*? By now, however, this is easy for us.

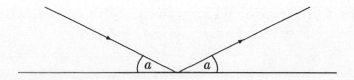

Figure 2-12: Light reflecting from a straight mirror

Just construct the line *tangent* to the curve, T, at the point of incidence, P. Then the angle of incidence is the angle between the *incoming* light ray and the *tangent*, and the angle of reflection is the angle between the *reflected* ray and the *tangent* (Figure 2-13). Simple?

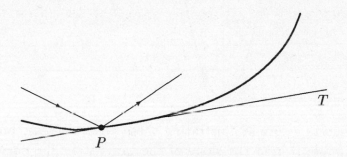

Figure 2-13: Light reflecting from a curved mirror

Now that we've seen another application of tangents we turn to our example, which involves a parabola. As we'll see in a moment, *parabolic mirrors* have important applications, which is why we have to understand them in detail. Parabolas are a class of curves, some of which have equations of the form $y = ax^2 + bx + c$, where $a, b,$ and c are constants. However, we're going to restrict ourselves in this example to a parabola with a particularly simple equation, $y = x^2$. Now, associated with every parabola is a point, F, called the *focus* (Figure 2-14). The focus is well-named because of the following property. Suppose we have a mirror in the shape of a parabola and a distant light source (such as the sun), whose rays coming in to the parabola are parallel. Then the reflections of the rays off the mirror are all focused at the point F. Conversely, if a light source such as a bulb is placed at the focus, then the rays emanating from the bulb reflect off the parabola in a *parallel* fashion. (This is the mathematics behind parabolic reflectors which capture solar energy and parabolic automobile headlights!)

Example 2.1   *What is the slope of the parabola whose equation is $y = x^2$ at the point $P = (2, 4)$ (Figure 2-15)?*

Solution: To find a point Q close to $P(2, 4)$ on the curve $y = x^2$, we choose a value of x close to 2, and compute the corresponding value of y from the equation $y = x^2$.

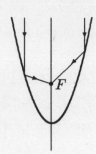

Figure 2-14: Focus of a parabola

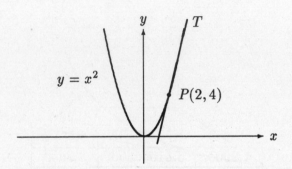

Figure 2-15: Slope of $y = x^2$ at (2,4)

For example, if we choose $x = 3$, then y will be equal to 9, so that the point Q has coordinates $(3, 9)$ (Figure 2-16). Then, from equation (2.3), the slope, S, of the secant

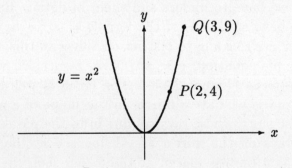

Figure 2-16: Initial choice of the point Q

line QP is given by

$$S = \frac{9 - 4}{3 - 2} = 5.$$

Next, we move Q closer to P along the curve, by choosing x equal to, say, 2.5. Since $2.5^2 = 6.25$, the new Q is $(2.5, 6.25)$. Again, from (2.3),

$$S = \frac{6.25 - 4}{2.5 - 2} = 4.5.$$

We continue in a similar fashion to obtain the following table:

x	y	S
3.0000	9.00000000	5.0000
2.5000	6.25000000	4.5000
2.1000	4.41000000	4.1000
2.0100	4.04010000	4.0100
2.0010	4.00400100	4.0010
2.0001	4.00040001	4.0001

Now let's take some points which approach 2 from the left.

x	y	S
1.0000	1.00000000	3.0000
1.5000	2.25000000	3.5000
1.9000	3.61000000	3.9000
1.9900	3.96010000	3.9900
1.9990	3.99600100	3.9990
1.9999	3.99960001	3.9999

It is clear from the tables that the slopes of the secant lines are approaching 4 as x approaches 2, which is therefore the slope of the tangent to the curve at the point P, thereby completing the solution.

Having computed the slope of the tangent, $m = 4$, however, we now turn to finding the equation of this line at $(2, 4)$. We use the point-slope formula, $y - y_0 = m(x - x_0)$, with $x_0 = 2, y_0 = 4$, and $m = 4$, to obtain $y - 4 = 4(x - 2)$ (or $y = 4x - 4$), as the equation of the tangent line.

Example 2.2   *Find the slope of the curve $y = x^2$ at the point $P = (x_0, y_0)$ (Figure 2-17).*

Solution: For a nearby point $Q = (x, y)$ the slope of the secant is

$$
\begin{aligned}
S \quad &= \quad \frac{y - y_0}{x - x_0} \\[2mm]
&= \quad \frac{x^2 - x_0{}^2}{x - x_0} \qquad (y = x^2; \ y_0 = x_0^2) \\[2mm]
&= \quad \frac{(x - x_0)(x + x_0)}{x - x_0} \qquad \textit{(factoring the numerator)} \\[2mm]
&= \quad x + x_0. \qquad \textit{(cancelling $x - x_0$)}
\end{aligned}
$$

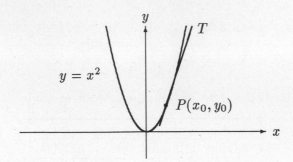

Figure 2-17: Slope of $y = x^2$ at (x_0, y_0)

Hence, for any point $Q = (x, y)$ on the curve, the slope of the secant line joining Q with P is

$$S = x + x_0. \tag{2.6}$$

As Q approaches P, it is clear that x approaches x_0. Hence, the limit of S as x approaches x_0 is equal to $2x_0$. For example, the slope at the point $(3, 9)$ is $2 \cdot 3 = 6$, while the slope at $(-5, 25)$ is $2 \cdot (-5) = -10$. We have thus found the slope at any point on the curve $y = x^2$.

We can now use this result to obtain the equation of the tangent line at any point (x_0, x_0^2). From the point-slope formula, $y - y_0 = m(x - x_0)$, so all we need do is substitute $m = 2x_0$, $y_0 = x_0^2$, obtaining $y = x_0^2 + 2x_0(x - x_0)$ or $y = 2x_0x - x_0^2$.

We mentioned earlier that our expectation that the approximation improves as x approaches x_0 is not always fulfilled. But how can it go wrong? The answer lies in the fact that the slope is not always defined. As we'll see in the following example, which involves the absolute value function, $f(x) = |x|$, curves that have *sharp corners* are among those for which the slope may fail to exist (at least at certain points). The absolute value function is important because $|a - b|$ measures the *distance* between the points a and b.

Example 2.3 *Find the slope at the origin of the curve with equation $y = |x|$ (Figure 2-18).*

Solution: Recall that $|x| = x$ if $x \geq 0$, but that $|x| = -x$ if $x < 0$. (Thus, for example, $|5| = 5, |-5| = -(-5) = 5$, and $|0| = 0$.) Since $y = |x|$, an alternate way of expressing this relationship is with a *split* formula:

$$y = \begin{cases} x & \text{if } x \geq 0 \\ -x & \text{if } x < 0. \end{cases}$$

Now both halves of this split formula represent straight lines: For $x \geq 0$, the graph of the equation $y = x$ is a line with slope 1 passing through the origin, while for $x < 0$, the

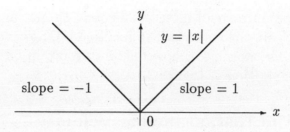

Figure 2-18: The slope of $y = |x|$ is undefined at $x = 0$

graph of $y = -x$ is a line with slope -1 which also passes through the origin. Putting these two lines together gives us the graph in Figure 2-18.

Let's try to compute the slope of this curve *at the origin*, $P = (0,0)$. First let Q approach P from the right, so that the coordinates of Q are (x, x). The slope is thus

$$\frac{y - y_0}{x - x_0} = \frac{x - 0}{x - 0} = 1.$$

Since the slope is 1 for *every* secant line, the right-hand limit of these slopes is also 1. However, when Q approaches P from the left, its coordinates are $(x, -x)$, which yields a slope of

$$\frac{y - y_0}{x - x_0} = \frac{-x - 0}{x - 0} = -1.$$

Again, the slope of every secant line is constant, but this time the constant is -1, producing a left-hand limit of -1. Finally, since the right- and left-hand limits do not agree, the slope at the origin *is not defined*. This is not the same as saying that the curve has slope 0 at the origin. For $y = |x|$, *no* number represents the slope at the origin. (Notice, also, that the slope does exist at every other point; for $x > 0$ the slope is 1 and for $x < 0$ the slope is -1.) The failure of this particular function to have a slope at 0 is *not* unusual. Any function which has a sharp corner fails to have a slope at that point. We see from this example that even for relatively simple functions such as $|x|$ the limiting process cannot be taken for granted.

The example we've just considered allows us to take a closer look at the problem of finding the slope of a curve. Suppose that the slope of a function f exists at a point x_0. What happens if we examine the graph of f in a small neighborhood of x_0 under a powerful microscope? In other words, what happens if we zoom in on the point x_0? (Most graphing calculators have a zoom button which allows you to magnify the graph. Those of you who own such a calculator can duplicate the steps we're about to perform.) Since f has a slope at x_0, it has a tangent at that point. We saw earlier in the case of a circle (Figures 2-5 and 2-6, page 13) that, as we zoom in, the tangent and the arc of the circle become almost indistinguishable. A similar phenomenon occurs for curves generally, which means that *in the small*, a curve that has a slope at a point x_0 is 'nearly a straight line' in a small interval surrounding x_0. In this situation, we say that the curve is 'locally linear' near x_0. The function $|x|$, however, is *not* locally linear in a vicinity of $x = 0$, for, no matter how much we zoom in on the function near the origin, all we see is a repetition of Figure 2-18.

2.1.2   Finding the Instantaneous Velocity

What we know: Average velocity of a moving body.

What we want to know: The velocity of a moving body at a particular instant (known as the *instantaneous velocity*).

How we do it: Approximate the instantaneous velocity by the average velocity over shorter and shorter time intervals.

Remember the familiar rate problems from high school algebra? (Perhaps you'd prefer to forget them!) They are all based upon the fundamental formula:

$$\text{distance} = \text{rate} \times \text{time} \quad (d = rt)$$

or

$$\text{rate} = \text{distance/time} \quad (r = d/t).$$

Thus, if a car travels at a rate $r = 30$ miles per hour for $t = 2$ hours, then the distance covered is $d = 30 \times 2 = 60$ miles. Similarly, if the car travels for $t = 2$ hours and covers $d = 100$ miles, then the rate, or *average velocity*, is

$$r = d/t = 100/2 = 50 \text{ miles per hour.}$$

Note the emphasis on the words 'average velocity.' Most cars travel at variable rates during a trip, but it is only an average velocity that we can calculate from the given information.

Now, without looking ahead, try to precisely answer the following:

Explain the meaning of a reading of 50 miles per hour on your speedometer.

The notion of average velocity is easy to comprehend and compute; it is nothing more than distance divided by time. So we *approximate the instantaneous velocity* by computing the average velocity of the car over say, one minute. But we may need to *refine* this approximation, which may not be very good, since this time period is not *very* short—there is plenty of room within it to change speed. Hence, to refine or improve the approximation, we take the average over a shorter time period, say 30 seconds. We continue to improve the approximation by taking shorter and shorter time periods, say 15 seconds, 10 seconds, 1 second, 0.1 second, and so forth. We anticipate better and better approximations as the time interval shrinks. But how do we obtain the precise value? We do this by continuing the process indefinitely, or, in mathematical parlance, by passing to the *limit*. More explicitly,

$\mathcal{L}$

> *instantaneous velocity is equal to the limit of average velocity, as the time interval shrinks to 0.*

Before introducing the details of the computations, take note once again of the 3-step process underlying the concept of instantaneous velocity: We began with an *approximation* of what we wanted; we *refined* the approximation; finally, we obtained the exact value by taking the *limit* $(\mathcal{A} - \mathcal{R} - \mathcal{L})$.

We're going to consider the case of motion along a straight line. While this restriction may seem artificial, there are situations in which it is realistic. For example, there *are* streets which are straight over long segments, and the motion of a bus or trolley traveling up and down such a street can be analyzed by the methods we are about to develop. Moreover, once we understand motion along a straight line, we can extend our study to the more usual case of motion along a curve, and the concepts of velocity and acceleration that we introduce here can be generalized to that situation. In this book, however, we will not discuss that generalization.

So suppose a body moves along a straight line, beginning at time t_0 at the point s_0, and ending at t_1, at which time it is at s_1 (Figure 2-19). Then the distance traveled,

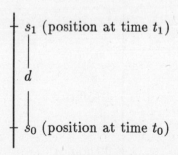

Figure 2-19: Motion along a straight line

d, is equal to $s_1 - s_0$, and the elapsed time, t, is equal to $t_1 - t_0$. Hence, the average velocity, r, is given by

$$r = \frac{s_1 - s_0}{t_1 - t_0}. \tag{2.7}$$

Now suppose that its *position* at any time, t, is given by the function $s = f(t)$.

Remark 2.2 It is *vital* to keep in mind that even though the motion of the body is along a *straight line*, the position function, f, which tells *where* on this straight line the body is located at any time, t, generally is *not* linear. Thus, the graph of $s = f(t)$ is usually not a straight line, but rather is a curve, which measures the distance between the body and some fixed *reference point* which corresponds to the *origin* on the s-axis (Figure 2-20). For example, the origin could represent your home, so that s measures

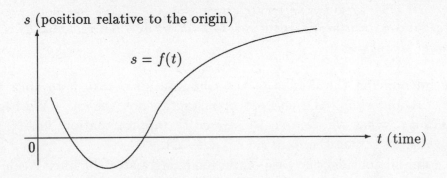

Figure 2-20: Graph of $s = f(t)$

how far you are from home. Let's suppose that your home, your office, a restaurant you like to eat at, and a stadium you attend all lie on a straight road. If you travel from your office, which is 5 miles north of your home to a baseball game at the stadium, which is 12 miles north of your home, then you have moved from $s = 5$ to $s = 12$, a distance of $12 - 5 = 7$ miles. If, however, you began your trip at the restaurant, 3

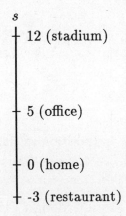

Figure 2-21: Total distance covered

miles *south* of your home (that is, $s = -3$, since your home corresponds to $s = 0$), and from there you drove to the stadium ($s = 12$), then your total distance would be $12 - (-3) = 15$ miles (Figure 2-21).

We want to compute the instantaneous velocity of the body at time, t_0. Choose a time t close to t_0, and suppose that the body is at position s at this time. Then, over the time interval from t_0 to t, the body has moved from s_0 to s (Figure 2-22). From

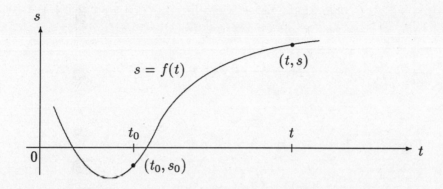

Figure 2-22: Initial approximation of the velocity

(2.7), the average velocity over this time interval is given by

$$r = \frac{s - s_0}{t - t_0}. \tag{2.8}$$

Since $s = f(t)$ and $s_0 = f(t_0)$, equation (2.8) can be rewritten as

$$r = \frac{f(t) - f(t_0)}{t - t_0}. \tag{2.9}$$

Here, $f(t) - f(t_0)$ represents the distance traveled in the time interval from t_0 to t, whose duration is $t - t_0$. We obtain the instantaneous velocity from (2.9) by letting the time interval shrink to 0. Hence, the instantaneous velocity at time t_0 is equal to

$$\lim_{t \to t_0} \frac{f(t) - f(t_0)}{t - t_0}. \tag{2.10}$$

Now, notice something interesting. Looking at Figure 2-23, we see that expression (2.8), which represents the average velocity from time t_0 to time t, also has a geometrical interpretation, namely, it is the *slope* of the line segment joining the points (t_0, s_0) and (t, s). In other words, it is the slope of the secant line joining these two points. What is striking about this observation is that the identical operation has arisen in what appear to be *totally different contexts*, and this is another clue that there is actually a close relationship between the two problems. In fact, compare expression (2.8) with equation (2.2), $S = (y - y_0)/(x - x_0)$, from Section 2.1.1, and also (2.10) with (2.5), which reads

$$m = \lim_{x \to x_0} \frac{f(x) - f(x_0)}{x - x_0}.$$

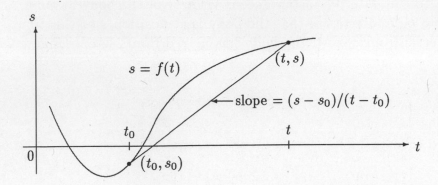

Figure 2-23: Geometric interpretation of average velocity

Except for the insignificant change of letters, each pair is identical in form. We see, when viewed properly, that the notions of *slope* and *velocity*, one *geometrical*, the other *physical*, are closely related. When expressions which are so similar occur in apparently different contexts, there is often a general principle underlying both of them. We shall soon see just what that principle is in this case.

Before leaving the subject of velocity, however, we briefly turn our attention to a related concept, acceleration. This topic is also familiar to us from our everyday experience with cars. In fact, the gas pedal of a car is often referred to as the accelerator. But just what is acceleration? It's nothing more than the change in velocity. Thus, if a car is moving at a constant velocity of 40 miles per hour, then its acceleration is *zero*. When a car begins to move, its velocity *increases*, so that its acceleration is *positive*. Conversely, when you hit the brakes, you *slow down* and your acceleration is *negative*.

How does calculus enter here? Well, we just said that acceleration is the change in velocity. Just as we defined instantaneous velocity to be the limit of the average velocity, so we define instantaneous acceleration to be the limit of the average acceleration. We'll pursue these ideas further in subsequent sections.

2.2   Definition of the Derivative

We now turn to the task of giving a name to the concept we have been discussing. If f is a function defined on an interval $[a, b]$, and if x_0 is a point in this interval, then the *derivative* of f at x_0 is defined as

$$\lim_{x \to x_0} \frac{f(x) - f(x_0)}{x - x_0} \qquad (2.11)$$

provided this limit exists. Notice that (2.11) is identical to (2.5) from Section 2.1.1 and differs from (2.10) only in the replacement of t by x.

For computational purposes, a slight variation of (2.11) is often more convenient. If we let $h = x - x_0$, the distance between x and x_0, then we have $x = x_0 + h$, and $x \to x_0$

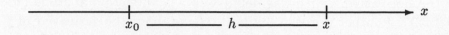

Figure 2-24: $h = x - x_0$

causes $h \to 0$ (Figure 2-24). Hence, we can rewrite (2.11) as

$$\lim_{h \to 0} \frac{f(x_0 + h) - f(x_0)}{h}. \tag{2.12}$$

Let's use the name we've given to our concept to describe the situations introduced in the previous section. Thus, the *slope* of a curve denoted by $y = f(x)$ at a point $P = (x_0, y_0)$ is equal to the *derivative* of the function f at x_0. The *instantaneous velocity* at time t_0 of a moving body whose position function is $s = f(t)$ is equal to the *derivative* of f at t_0. The *instantaneous acceleration* of the same body is equal to the *derivative* of the velocity function and, hence, is the *second derivative* of the position function, f. (We'll have more to say about the second derivative in subsequent sections.)

Referring back to our discussion at the end of Example 2.3 (page 21), we are reminded that if a function f is differentiable at a point x_0, then f is locally linear near x_0. This means that a small segment of the graph of f in the vicinity of x_0 is very nearly a straight line.

2.3   Notation for the Derivative

Our new concept needs a notation. For the derivative there are a number of different symbols in common use, and while this can cause some initial confusion, it actually reflects the richness of the concept, since each notation emphasizes another of its aspects. We introduce here two of the most popular notations for the derivative, and discuss some of the advantages and disadvantages of each.

The most common notation for the derivative of f at the point x_0 is $f'(x_0)$, (read "f prime of x_0") which has the advantage of clarity in two respects: Both the function, f, and point, x_0, are specified.

Example 2.4   *Find the derivative of the function $f(x) = x^2$ at the point $x = x_0$.*

Solution: We saw earlier (Example 2.1) that the slope of the curve $y = x^2$ at the point $x_0 = 2$ is 4. Since $f(x) = x^2$, we have $f'(2) = 4$. Similarly, since the slope at a general point x_0 was found to be $2x_0$ (Example 2.2), we have $f'(x_0) = 2x_0$.

Remark 2.3   Notice that since the point x_0 is completely general, we can simplify the notation somewhat by dropping the subscript and writing $f'(x) = 2x$. However, some caution is necessary in doing so, because it is possible to lose sight of the original meaning of the derivative, which was defined on a point-by-point basis (see equations (2.10) and (2.12)). In other words, we initially spoke about the derivative of f *at a point*, x_0, while now we are referring to the derivative of the *function* f. The danger lies in thinking of the operation of differentiation as nothing more than a formal manipulation of symbols; i.e., of simply saying that the derivative of $f(x) = x^2$ is $2x$. There is no harm done in this, since we can recover the value of the derivative at a specific point by substitution. All too often, however, students quickly forget the original significance of the derivative and concentrate, instead, on the manipulative aspects. *So be forewarned!*

Under this more general approach to the derivative, equation (2.12) takes the form

$$f'(x) = \lim_{h \to 0} \frac{f(x+h) - f(x)}{h}. \tag{2.13}$$

A second, commonly used notation for the derivative requires a switch of emphasis from functions to variables. We write $y = f(x)$, and return to formula (2.2) in Section 2.1.1, in which we computed

$$S = \frac{y - y_0}{x - x_0}. \tag{2.14}$$

Earlier we denoted the quantity $x - x_0$ by Δx and $y - y_0$ by Δy. Then $S = \Delta y / \Delta x$, and since the slope of the tangent line, m, is obtained by taking the limit of $(y - y_0)/(x - x_0)$ as $x \to x_0$, we have

$$
\begin{aligned}
m &= \lim_{x \to x_0} \frac{y - y_0}{x - x_0} \\
&= \lim_{\Delta x \to 0} \frac{\Delta y}{\Delta x}.
\end{aligned} \tag{2.15}
$$

The use of deltas leads naturally to the following notation for the derivative:

$$\frac{dy}{dx} = \lim_{\Delta x \to 0} \frac{\Delta y}{\Delta x}. \tag{2.16}$$

This notation emphasizes a key aspect of the derivative, namely, it represents the *rate of change* of y with respect to x. On the other hand, a serious deficiency in the notation is that there is no place to indicate the point at which the derivative is sought.

Example 2.5 *Find dy/dx for $y = x^2$.*

Solution: For $y = x^2$, we have $dy/dx = 2x$. This is fine when we are looking for the derivative of the *function* $y = x^2$. But how would we indicate, say, the derivative of this function when $x = 2$? The most common way of doing this is to write

$$\left. \frac{dy}{dx} \right|_{x=2} = 2x \big|_{x=2} = 4.$$

Of course, this is clumsier than writing $f'(2)$. So why do we have two separate notations for the same concept? One answer is historical: Different mathematicians used different notations, and no single one has achieved universal acceptance. But a second reason is that each notation has its own advantages in different situations, and it is therefore a good idea to become comfortable with both notations and to recognize where each is better used. We will discuss this later on in the chapter.

Other notations for the derivative include $\dot{y}$ (read "y-dot"; this is Newton's notation, but is rarely seen today), $df/dx, Df, D_x y$, and $D_x f$, among others.

Now that we have a notation for the derivative, we can obtain a general formula for the equation of the tangent to a curve $y = f(x)$, at a point $(x_0, f(x_0))$. Using the point-slope formula, $y - y_0 = m(x - x_0)$, we substitute $y_0 = f(x_0)$ and $m = f'(x_0)$ (since the derivative at x_0 is the same as the slope of the tangent line), yielding

$$y - f(x_0) = f'(x_0)(x - x_0)$$

or

$$y = f(x_0) + f'(x_0)(x - x_0). \tag{2.17}$$

How do we denote the second derivative, which we introduced at the end of the last section? For the prime notation, we write $f''(x)$. Thus, if $f(x) = x^2$, then $f'(x) = 2x$ and $f''(x) = 2$. The alternate notation for the second derivative is

$$\frac{d^2 y}{dx^2}.$$

So for the same example, if $y = x^2$, then

$$\frac{dy}{dx} = 2x$$

$$\frac{d^2 y}{dx^2} = 2.$$

2.4   Calculating Derivatives

When we try to apply the definition of the derivative given in Section 2.2 to actual computations, we immediately encounter difficulties of an algebraic nature, which do not preclude our ability to perform the calculations, but certainly encourage us to seek alternate methods. In the next example, look at how much work is needed to compute the derivative of a relatively simple function.

Example 2.6 *Find $k'(x)$ for $k(x) = (3x - 1)/(x + 4)$.*

Solution: From (2.13),

$$k'(x) = \lim_{h \to 0} \frac{k(x+h) - k(x)}{h}$$

$$= \lim_{h \to 0} \frac{[3(x+h) - 1)/((x+h)+4] - [(3x-1)/(x+4)]}{h}$$

$$= \lim_{h \to 0} \frac{(3x+3h-1)(x+4) - (3x-1)(x+h+4)}{(x+h+4)(x+4)h} \quad \text{(combining)}$$

$$= \lim_{h \to 0} \frac{3x^2 + 12x + 3hx + 12h - x - 4 - 3x^2 - 3hx - 12x + x + h + 4}{(x+h+4)(x+4)h}$$

$$= \lim_{h \to 0} \frac{13h}{(x+h+4)(x+4)h} \qquad \text{(cancelling in numerator)}$$

$$= \lim_{h \to 0} \frac{13}{(x+h+4)(x+4)} \qquad \text{(cancelling h)}$$

$$= \frac{13}{(x+4)^2}. \qquad \text{(since } h \to 0\text{)}$$

So the *mechanical* aspects of differentiation are complex (and tedious!). We desperately need shortcuts for computing derivatives, which avoid using the formal definition. To accomplish this, we will consider the various ways in which two functions may be combined to form a new function. There are practical reasons for studying such combinations. For example, if $R(x)$ is the *revenue* (income) a company receives if it sells x cars and $C(x)$ is the *cost* of manufacturing them, then $P(x) = R(x) - C(x)$ is the *profit* it makes on these cars. Hence, *subtraction* is one method of combining functions. *Division* is another, as we saw in Example 2.6, where the function $(3x - 1)$ is divided by the function $(x + 4)$. Other ways of combining two functions f and g include addition $(f + g)$, multiplication (fg) and composition $(f \circ g)$. The last one means the following: $(f \circ g)(x) = f(g(x))$. The techniques for handling such combinations are generally known as *rules* of differentiation, but are actually *theorems*, whose proofs are found in all standard calculus texts. There are a number of such rules, the most important of them being the following:

Constant Multiple Rule: The derivative of cf (where c is a constant) is cf'.

Sum Rule: The derivative of $f + g$ is $f' + g'$.

Difference Rule: The derivative of $f - g$ is $f' - g'$.

Product Rule: The derivative of fg is $f'g + fg'$.

Quotient Rule: The derivative of f/g is

$$\frac{f'g - fg'}{g^2}.$$

Chain Rule: The derivative of the composite function $f \circ g$ is $(f' \circ g) \cdot g'$ (so that $(f \circ g)'(x) = f'(g(x)) \cdot g'(x)$).

Power Rule: The derivative of x^n is nx^{n-1}.

The last of these rules, which applies to a specific class of functions, is somewhat different from the first six, which may be used for *any* differentiable functions f and g. The power functions are important enough, however, to accord them a special rule. For reference, we include a table of derivatives (page 32) of most of the basic functions that you'll encounter in a calculus course. These specific results, together with the general rules outlined above, allow us to compute derivatives of all sorts of complicated functions which are constructed from these basic ones. It may seem incredible, but the only functions you'll see in this course for which derivatives are calculated *using the definition* are x^n, $\sin x$, e^x, and $\ln x$. The derivatives of *all* the others are computed mainly by using the rules introduced above. This is possible because more complicated functions are built up from these basic ones algebraically. For example, every polynomial is obtained by adding or subtracting constant multiples of the power functions, x^n. We can thus compute the derivative of any polynomial through use of the Constant Multiple, Sum, Difference, and Power Rules. Similarly, rational functions are quotients of two polynomials, so their derivatives may be calculated by applying the Quotient Rule to the polynomials. The derivative of $\cos x$ is obtained from that of $\sin x$ through a trigonometric identity, and the other trig functions are simple combinations of $\sin x$ and/or $\cos x$.

Example 2.7 *Let* $f(x) = 3x - 1$, $g(x) = x + 4$, *and* $c = 7$. *Then* $f'(x) = 3$ *and* $g'(x) = 1$, *so that the derivative of*

a. cf is $7 \cdot 3$;

b. $f + g$ is 4;

c. $f - g$ is 2;

d. fg is $3 \cdot (x + 4) + (3x - 1) \cdot 1 = 6x + 11$;

e. f/g is $(3 \cdot (x + 4) - (3x - 1) \cdot 1)/(x + 4)^2 = 13/(x + 4)^2$;

f. $f \circ g$ is $3 \cdot 1 = 3$.

(**e.** is confirmed by our earlier, lengthy calculation.)

Example 2.8 *Find the equation of the line tangent to the curve* $y = x^3 - 4x^2 + 4x + 1$ *at the point* $(1, 2)$.

Solution: Letting $f(x) = x^3 - 4x^2 + 4x + 1$, we obtain $f'(x) = 3x^2 - 8x + 4$ from the Constant Multiple, Sum, and Power Rules, so that $f'(1) = -1$. Hence, from (2.17), the equation of the tangent line is $y = 2 - (x - 1)$ or $y = -x + 3$ (Figure 2-25).

Table of Derivatives

f	**f′**	**f**	**f′**
$\sin x$	$\cos x$	$\sin(g(x))$	$\cos(g(x))g'(x)$
$\cos x$	$-\sin x$	$\cos(g(x))$	$-\sin(g(x))g'(x)$
$\tan x$	$\sec^2 x$	$\tan(g(x))$	$\sec^2(g(x))g'(x)$
$\cot x$	$-\csc^2 x$	$\cot(g(x))$	$-\csc^2(g(x))g'(x)$
$\sec x$	$\sec x \tan x$	$\sec(g(x))$	$\sec(g(x))\tan(g(x))g'(x)$
$\csc x$	$-\csc x \cot x$	$\csc(g(x))$	$-\csc(g(x))\cot(g(x))g'(x)$
e^x	e^x	$e^{g(x)}$	$e^{g(x)}g'(x)$
a^x	$a^x \ln a$	$a^{g(x)}$	$a^{g(x)}g'(x)\ln a$
$\ln x$	$\dfrac{1}{x}$	$\ln(g(x))$	$\dfrac{g'(x)}{g(x)}$
$\sin^{-1} x$	$\dfrac{1}{\sqrt{1-x^2}}$	$\sin^{-1}(g(x))$	$\dfrac{g'(x)}{\sqrt{1-(g(x))^2}}$
$\cos^{-1} x$	$\dfrac{-1}{\sqrt{1-x^2}}$	$\cos^{-1}(g(x))$	$\dfrac{-g'(x)}{\sqrt{1-(g(x))^2}}$
$\tan^{-1} x$	$\dfrac{1}{1+x^2}$	$\tan^{-1}(g(x))$	$\dfrac{g'(x)}{1+(g(x))^2}$
$\cot^{-1} x$	$\dfrac{-1}{1+x^2}$	$\cot^{-1}(g(x))$	$\dfrac{-g'(x)}{1+(g(x))^2}$
$\sec^{-1} x$	$\dfrac{1}{x\sqrt{x^2-1}}$	$\sec^{-1}(g(x))$	$\dfrac{g'(x)}{g(x)\sqrt{(g(x))^2-1}}$
$\csc^{-1} x$	$\dfrac{-1}{x\sqrt{x^2-1}}$	$\csc^{-1}(g(x))$	$\dfrac{-g'(x)}{g(x)\sqrt{(g(x))^2-1}}$

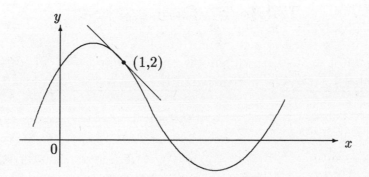

Figure 2-25: Tangent to $y = x^3 - 4x^2 + 4x + 1$ at (1,2)

Example 2.9 *Let $f(x) = x^2$ and $g(x) = \sin x$. Find the derivatives of the various combinations of f and g, as in Example 2.7.*

Solution: Here, $f'(x) = 2x$ and $g'(x) = \cos x$, yielding the following table of the derivatives of various combinations of f and g.

Function	Derivative	Reason
$x^2 + \sin x$	$2x + \cos x$	Sum Rule
$x^2 - \sin x$	$2x - \cos x$	Difference Rule
$x^2 \sin x$	$2x \sin x + x^2 \cos x$	Product Rule
$x^2 / \sin x$	$(2x \sin x - x^2 \cos x)/ \sin^2 x$	Quotient Rule
$\sin x / x^2$	$(x^2 \cos x - 2x \sin x)/x^4$ $= (x \cos x - 2 \sin x)/x^3$	Quotient Rule
$\sin(x^2)$	$\cos(x^2) \cdot 2x$	Chain Rule
$(\sin x)^2$	$2 \sin x \cos x$	Chain Rule

The last two results perhaps need some elaboration. We obtain $\sin(x^2)$ by the composition of the functions f and g in the order $g \circ f = \sin(f) = \sin(x^2)$, while $f \circ g = g^2 = (\sin x)^2$. Now, by the Chain Rule, the derivative of $g \circ f$ is

$$(g' \circ f) \cdot f' = \cos(x^2) \cdot 2x,$$

while the derivative of $f \circ g$ is

$$(f' \circ g) \cdot g' = 2(\sin x) \cdot \cos x.$$

Exercise 2.1 *Let $f(x) = x^4$ and $g(x) = \cos x$. Find the derivatives of all the combinations as in Example 2.9.*

Example 2.10 *Find the derivative of $x^3 + \tan x/x$.*

Solution: Let $f(x) = x^3$ and $g(x) = \tan x / x$. The derivative we want is thus the derivative of the function $f + g$, which is $f' + g'$ by the Sum Rule. Now $f'(x) = 3x^2$ (by the Power Rule), but g itself is a combination of two functions, $\tan x$ and x. So we let $h(x) = \tan x$ and $k(x) = x$. With this notation, g is equal to h/k, so that the Quotient Rule may be applied. Since the derivative of $\tan x$ is $\sec^2 x$, we obtain

$$
\begin{aligned}
g'(x) &= \frac{h'(x)k(x) - h(x)k(x)'}{k^2(x)} \\
&= \frac{\sec^2 x \cdot x - \tan x}{x^2}.
\end{aligned}
$$

Hence,

$$
\begin{aligned}
(f(x) + g(x))' &= f'(x) + g'(x) \\
&= 3x^2 + \frac{\sec^2 x \cdot x - \tan x}{x^2} \\
&= \frac{3x^4 + x \sec^2 x - \tan x}{x^2}.
\end{aligned}
$$

Example 2.11 *Find the first and second derivatives of $f(x) = x^2 \sin x$.*

Solution: Using the Product Rule, we obtain

$$
f'(x) = 2x \sin x + x^2 \cos x
$$

and

$$
f''(x) = 2 \sin x + 2x \cos x + 2x \cos x - x^2 \sin x = 2 \sin x + 4x \cos x - x^2 \sin x.
$$

Remark 2.4 An analysis of the Chain Rule enables us to examine in closer detail the advantages of the alternate notation for the derivative. Suppose $y = f(g(x))$. The Chain Rule states that $dy/dx = (f' \circ g) \cdot g'$, but this is a mixture of the two notations. How will this rule appear in the alternate notation? Recall that the derivative of $f \circ g$ at the point x is $f'(g(x)) \cdot g'(x)$. Now introduce an intermediate variable, $u = g(x)$. Then we have $y = f(u)$, where $u = g(x)$. Hence, $f'(g(x)) = f'(u) = dy/du$, and $g'(x) = du/dx$, so that we obtain

$$
\frac{dy}{dx} = \frac{dy}{du}\frac{du}{dx}. \tag{2.18}
$$

This approach also helps us understand the origin of the name 'Chain Rule': The variable u is the link in the chain between the variables x and y.

Example 2.12 *Find the derivative of $\sqrt{1 + x^2}$.*

Solution: Let $y = \sqrt{1 + x^2}$, and let $u = 1 + x^2$. Then, $y = \sqrt{u} = u^{1/2}$ and $u = 1 + x^2$. Hence, from (2.18), we obtain

$$
\begin{aligned}
\frac{dy}{dx} &= \frac{dy}{du}\frac{du}{dx} \\
&= (1/2)u^{-1/2} \cdot (2x) \\
&= x(1 + x^2)^{-1/2}.
\end{aligned}
$$

Exercise 2.2 *Find the derivative of* $(\sin x + x \cos x)^{1/2}$.

The Chain Rule also gives us an opportunity to elaborate on the notion of *rate of change*, which we've seen is an important aspect of the derivative. *dy/dx* measures the rate at which the variable y is changing with respect to the variable x. The following examples drawn from various fields illustrate how important and widespread this concept is.

- If $P = f(t)$ is the *profit* of a company in the *year t*, then dP/dt measures how fast the company's profits are changing with respect to time. (Hopefully, the change is *positive*, indicating growing profits!)

- If $P = g(x)$ is the *profit* of a company when it *produces x items*, then dP/dx is known in economics as the *marginal profit*, that is, the additional profit (or loss, if dP/dx is negative) that comes from producing one more item when the production level is currently x units.

- If $R = f(t)$ is the *amount* of radioactive substance in a landfill at *time t*, then dR/dt represents the rate at which this substance is increasing (if new radioactive material is added), or decreasing (if no new material is added, so that the substance is allowed to decay).

- If $v = f(t)$ is the *velocity* at *time t* of a car moving in a straight line, then dv/dt is the change in velocity, or the *acceleration* of the car. If dv/dt is positive, then the car is speeding up, while if it is negative, then the car is slowing down. (What's going on if $dv/dt = 0$?)

- In chemistry, the *reaction rate* is the *rate of change* of concentration of a chemical substance.

- In the study of electricity, *current* is the *rate of change* of the amount of *electrical charge* as a function of time.

- Finally, perhaps the best known of all, is the *inflation rate*, which measures the *change in prices*.

The interpretation of the derivative as a rate of change leads to a simple, but easily accessible, understanding of the Chain Rule. Suppose we have three variables, y, u, and x, with y a function of u, and u a function of x. Then y is indirectly a function of x. Now, suppose that at a certain point, y is changing 3 times as fast as u and u is changing 5 times as fast as x. Then, clearly, y is changing $3 \cdot 5 = 15$ times as fast as x. *But this is just the Chain Rule!* For the rate of change of y with respect to u is given by dy/du and of u with respect to x by du/dx. Hence, $dy/dx = (dy/du)(du/dx) = 3 \cdot 5 = 15$.

2.5   Applications of the Derivative

The examples just introduced give us a good indication of the widespread applicability of the derivative in the sciences, engineering, medicine, and the social sciences. You will find many additional applications in your text, including the solving of maximum-minimum and motion problems, the graphing of functions, related rates, and many others. We will consider two important *mathematical* applications of the derivative. The first is a powerful method for solving equations, developed by Isaac Newton. The second, known as Taylor polynomials, leads to efficient numerical computation of many functions, such as the trigonometric, exponential, and logarithmic functions. Because of the length of the exposition of these topics, we devote a separate chapter to these applications.

Solved Problems

Before introducing the solved problems, it will be useful to discuss some additional aspects of the derivative. As usual, you can consult your text if you feel you need more details, but the presentation will be adequate for solving the problems which follow.

- The derivative conveys useful information about the graphs of functions. Recall that a line which *rises* has a *positive* slope and vice versa. While the slope of a curve generally changes from point to point, a differentiable function rises over an *interval* $[a, b]$ if and only if it has a positive *derivative* on $[a, b]$. Similarly, a function *decreasing* on an interval has a *negative* derivative there. If we know where the derivative of a function is positive and where it is negative, then we will know where its graph is increasing and where it is decreasing. Conversely, knowledge of the rise and fall of the function (say we have been presented with the graph of the function) yields important information about its derivative.

- An important consideration in the topic of graphing is the curvature of a function, or how it bends. A curve such as the parabola $y = x^2$ is said to be *concave up*;

the 'inverted parabola,' $y = -x^2$, is *concave down*. Calculus enters here because of the connection between the concavity of a function and its *second derivative*: A function, f, is concave up on an interval $[a, b]$ if and only if $f''(x) \geq 0$ for all x in $[a, b]$. However, many, if not most, functions that we'll encounter *change concavity*; they are concave up on an interval, but then concave down on an adjacent one. We can use the second derivative to determine the concavity of the function.

• Another fact to take note of is that not every function is given by a formula. Situations exist in which the *graph* of a function is presented, and we must then answer questions about its behavior and that of its derivative. We will see problems of this type in this section.

2.1 Find the slope of the curve $y = 2x^3$ at the point $P = (4, 128)$, directly from the definition.

Solution: If we choose $Q = (5, 250)$, then the slope S of the secant joining P and Q is

$$S = \frac{250 - 128}{5 - 4} = 122.$$

Choosing a sequence of points approaching P yields the following table.

x	y	S
5.000	250.000000	122.00
4.500	182.250000	108.50
4.100	137.842000	98.842
4.010	128.962402	96.240
4.001	128.096024	96.024

The values of S are approaching 96 as x approaches 4 from the right. (Compute the values of S for $x = 4.0001$ and 4.00001 to convince yourself of this.) A similar calculation from the left (which you should perform) confirms this result. Hence the slope of the curve $y = 2x^3$ at $(4, 128)$ is 96.

2.2 Find the derivative of $f(x) = \sqrt{1 + x}$ at the point $x = 3$ directly from the definition of the derivative. Then check your answer using the rules of differentiation.

Solution:

$$f'(3) = \lim_{h \to 0} \frac{f(3 + h) - f(3)}{h}$$

$$= \lim_{h \to 0} \frac{\sqrt{4 + h} - 2}{h}$$

$$= \lim_{h\to 0} \frac{(\sqrt{4+h}-2)(\sqrt{4+h}+2)}{h(\sqrt{4+h}+2)}$$

$$= \lim_{h\to 0} \frac{4+h-4}{h(\sqrt{4+h}+2)}$$

$$= \lim_{h\to 0} \frac{h}{h(\sqrt{4+h}+2)}$$

$$= \lim_{h\to 0} \frac{1}{\sqrt{4+h}+2}$$

$$= \frac{1}{4}.$$

Using the chain rule, we obtain

$$y = \sqrt{u}, \quad \text{where} \quad u = 1 + x.$$

Hence,

$$\frac{dy}{dx} = \frac{dy}{du}\frac{du}{dx} = \frac{1}{2}\, u^{-\frac{1}{2}}\,(1) = \frac{1}{2\sqrt{u}} = \frac{1}{2\sqrt{1+x}}.$$

Now setting $x = 3$, we obtain $1/(2\sqrt{4}) = 1/4$.

2.3   Prove the Sum Rule for differentiation: The derivative of $f + g$ is $f' + g'$.

Solution: By the definition of the derivative, to find the derivative of $f + g$ we must evaluate the following limit:

$$\lim_{h\to 0} \frac{[f(x+h)+g(x+h)] - [f(x)+g(x)]}{h}$$

$$= \lim_{h\to 0} \frac{[f(x+h)-f(x)] + [g(x+h)-g(x)]}{h}$$

$$= \lim_{h\to 0} \left[\frac{f(x+h)-f(x)}{h} + \frac{g(x+h)-g(x)}{h}\right]$$

$$= \lim_{h\to 0} \frac{f(x+h)-f(x)}{h} + \lim_{h\to 0} \frac{g(x+h)-g(x)}{h}$$

$$= f'(x) + g'(x).$$

2.4   Find f' and f'' for the following functions:

   a. $f(x) = x^4 + 2x^3 - x + 5$

   b. $f(x) = e^{x^2}$

Solution:

a. $f'(x) = 4x^3 + 6x^2 - 1$ and $f''(x) = 12x^2 + 12x$.

b. f is a composite function, so we'll need the Chain Rule. We first rewrite the function in variable form (which is more convenient when using the Chain Rule), obtaining

$$y = e^{x^2}.$$

We apply the Chain Rule by introducing an intermediate variable, u, as follows:

$$y = e^u, \quad \text{where} \quad u = x^2.$$

Now,

$$\frac{dy}{dx} = \frac{dy}{du}\frac{du}{dx} = e^u \cdot x^2 = e^{x^2} \cdot 2x.$$

We now use the Product Rule to find the second derivative. Letting $g(x) = e^{x^2}$ and $h(x) = 2x$, we have

$$
\begin{aligned}
(g(x)h(x))' &= g'(x)h(x) + g(x)h'(x) \\
&= e^{x^2}(2x) \cdot (2x) + e^{x^2} \cdot 2 \\
&= e^{x^2}(4x^2 + 2),
\end{aligned}
$$

which is equal to $f''(x)$.

2.5   At which points does the function defined by the split formula

$$f(x) = \begin{cases} 2x - 1 & x < 0 \\ x - 1 & 0 \leq x \leq 2 \\ 3x - 5 & 2 < x \end{cases}$$

fail to have a derivative? Sketch the graph of the function.

Solution: When we investigated the function $|x|$ at the origin (page 20), we saw that a function which has a sharp corner has no derivative at that corner. In this problem, f has sharp corners at $x = 0$ and 2. At $x = 0$ the slope changes from 2 to 1, while at $x = 2$ it changes from 1 to 3. The sketch of the graph is left to the reader.

2.6   Suppose $f(0) = 3$, $f'(0) = -2$, $g(0) = 7$, $g'(0) = 4$. Let $h = fg$ and $k = f/g$. Find $h'(0)$ and $k'(0)$.

Solution: By the product rule,

$$h'(0) = f'(0)g(0) + f(0)g'(0) = (-2) \cdot 7 + 3 \cdot 4 = -2.$$

Similarly, using the quotient rule, we obtain

$$k'(0) = \frac{f'(0)g(0) - f(0)g'(0)}{(g(0))^2} = \frac{(-2) \cdot 7 - 3 \cdot 4}{49} = -\frac{26}{49}.$$

2.7 Suppose f and g are differentiable functions with $g(0) = -2$, $g'(0) = 2$ and $f'(-2) = 1/2$. Let $h = f \circ g$. Find $h'(0)$.

Solution: Since $h = f \circ g$, we need the chain rule to find h'. Hence, $h'(0) = f'(g(0)) \cdot g'(0) = f'(-2) \cdot 2 = (1/2) \cdot 2 = 1$.

2.8 f is a function whose *derivative* is $\sin(x^2)$. (Don't try to figure out what f is — it's not an elementary function.) Find the derivative of $f(\ln x)$.

Solution: Let $y = f(\ln x)$ and $u = \ln x$. Then

$$y = f(u), \quad u = \ln x.$$

Apply the chain rule:

$$\frac{dy}{dx} = \frac{dy}{du}\frac{du}{dx} = f'(u)\frac{1}{x} = \sin((\ln x)^2)\frac{1}{x}.$$

2.9 Find dy/dx and express it solely in terms of x:

$$y = \sqrt{u}, \quad u = v^3 + 1, \quad v = \sin x.$$

Solution:

$$\begin{aligned}
\frac{dy}{dx} &= \frac{dy}{du}\frac{du}{dv}\frac{dv}{dx} \\[2mm]
&= \left(\frac{1}{2}u^{-1/2}\right)(3v^2)(\cos x) \\[2mm]
&= \left(\frac{1}{2}u^{-1/2}\right)(3\sin^2 x)(\cos x) \\[2mm]
&= \frac{3}{2}\frac{\sin^2 x \cos x}{\sqrt{\sin^3 x + 1}}.
\end{aligned}$$

2.10 Figure 2-26 contains the graph of a function given by the equation $f(x) = ax^3 + bx + c$. What are the values of a, b, and c?

Solution: First note that $f(0) = 0$. Since $f(0) = c$, we see that $c = 0$, thereby reducing $f(x)$ to $ax^3 + bx$. We need 2 additional conditions to determine a and b, and we get them from the fact that $f(1) = -1$ and $f'(1) = 0$. Thus,

$$f(1) = a + b = -1 \quad \text{and} \quad f'(1) = 3a + b = 0.$$

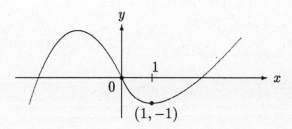

Figure 2-26: Graph of $y = ax^3 + bx + c$

The solution to this simultaneous system of equations is $a = 1/2$, $b = -3/2$, so that $f(x) = x^3/2 - 3x/2$ is the desired function.

2.11   The line T is tangent to the curve $y = f(x)$ in Figure 2-27.

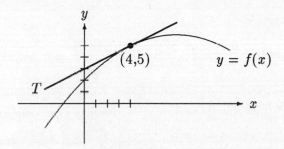

Figure 2-27: A curve and its tangent

a.   Find $f(0)$.
b.   Find $f'(4)$.

Solution:

a. $f(0)$ can be read off the graph: It's the place where the *curve* crosses the y-axis, which occurs at $y = 2$. Hence, $f(0) = 2$.

b.   The derivative $f'(4)$ is equal to the slope of T, the line tangent to the curve $y = f(x)$ at $x = 4$. The slope of $T = (5 - 3)/(4 - 0) = 1/2$.

2.12   Show that the graph of the function $f(x) = x^3 + x - 4$ is always increasing.

Solution: We first calculate the derivative of f, obtaining $f'(x) = 3x^2 + 1$, which is always *positive*. Hence, by our earlier remarks, the graph of f increases for all x.

2.13   Determine where the graph of the function f is increasing and where it is decreasing for each of the following:

a. $f(x) = 2x^3 - 3x^2 - 12x + 3$

b. $f(x) = 2\sin x - x$, for $x \in [0, 2\pi]$.

Solution:

a. $f'(x) = 6x^2 - 6x - 12 = 6(x+1)(x-2)$. This quantity is ≤ 0 for $-1 \leq x \leq 2$, and positive otherwise. Hence, the graph of f is falling for $-1 \leq x \leq 2$ and rising elsewhere.

b. $f'(x) = 2\cos x - 1$. In the interval $[0, 2\pi]$, $\cos x \geq 1/2$ for $0 \leq x \leq \pi/3$ $(\pi/3 = 60°)$ and for $5\pi/3 \leq x \leq 2\pi$, and is $< 1/2$ for all other values. Thus, the graph of f is increasing for $x \in [0, \pi/3] \cup [5\pi/3, 2\pi]$, and decreasing in $[\pi/3, 5\pi/3]$ (Figure 2-28).

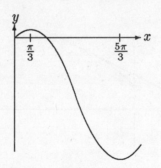

Figure 2-28: Where $f(x) = 2\sin x - x$ increases and decreases

2.14   Show that the function $f(x) = x + \cos x$ has *no* local maximum or minimum points, even though its derivative is 0 infinitely often.

Solution: $f'(x) = 1 - \sin x$, which is 0 whenever $\sin x = 1$. This occurs at $x = \pi/2 + 2k\pi$, where $k = 0, \pm1, \pm2, \ldots$, so that there are an infinite number of such points. However, $f'(x) > 0$ for all other x, so that the curve is constantly rising. (It will be instructive for you to draw a sketch of the curve.)

2.15   **True or False:** If f and g are both increasing functions on $[a, b]$, then fg must also increase on $[a, b]$?

Solution: False! Let $h = fg$. Then, by the product rule, $h' = f'g + fg'$. If f and g are both *negative* increasing functions, then h' will also be negative on $[a, b]$, and so will be decreasing. An example of this is $f(x) = g(x) = \sin x$ on $[-\pi/2, 0]$. The product, $fg = \sin^2 x$, which decreases from 1 to 0 on this interval.

2.16 A mold has a mass of $3t^2$ grams after t hours of growth.

    a.   How *much* does it grow during the time interval [3,3.01]?

    b.   What is its *average rate of growth* during this time interval?

    c.   What is its *instantaneous rate of growth* when $t = 3$?

Solution:

    a.   At $t = 3$ hours the mold weighs 27 grams, while at $t = 3.01$ hours its weight is 27.1803 grams, so it has grown by .1803 grams in this time interval.

    b.   The average rate of growth is $.1803/.01 = 18.03$ grams per hour.

    c.   To find the instantaneous rate of growth when $t = 3$, we calculate the derivative of $3t^2$ and evaluate it at $t = 3$. The derivative is $6t$ which equals 18 when $t = 3$.

2.17 A point is moving along the curve $y = \sqrt{x}$ in such a way that its x-coordinate is increasing at the rate of 4 feet per second.

    a.   At what rate is its *y-coordinate* changing when $x = 4$?

    b.   At what rate is its *slope* changing when $x = 4$?

Solution:

    a.   We are told that $dx/dt = 4$, and we want to find dy/dt. So we use the chain rule:

$$\frac{dy}{dt} = \frac{dy}{dx}\frac{dx}{dt} = \frac{1}{2\sqrt{x}}\frac{dx}{dt} = \frac{2}{\sqrt{x}}.$$

Thus, when $x = 4$, $dy/dt = 1$.

    b.   The slope of the curve is given by dy/dx. Call this quantity z : $z = dy/dx$. Again, by the chain rule,

$$\frac{dz}{dt} = \frac{dz}{dx}\frac{dx}{dt} = -\frac{1}{4}x^{-3/2}\frac{dx}{dt} = -x^{-3/2}.$$

At $x = 4$, $\frac{dz}{dx} = -(4)^{-3/2} = -\frac{1}{8}$. Looking at the graph of $y = \sqrt{x}$ (Figure 3-6, page 63), we see that this makes sense: As x moves to the right, y *increases* (that's why $dy/dt > 0$), but the slope *decreases*.

2.18 Consider the following table, which involves the Consumer Price Index (CPI), a measure of inflation:

Month	CPI	Change in Index	% Change in Index
April	400	—	—
May	402	2	.5 %
June	403	1	.25 %

Are the following statements correct? Explain.

a. Prices rose in June.

b. Inflation rose in June.

Solution:

a. Since the CPI is higher in June than in May, the statement is correct.

b. The inflation *rate* is a measure of *how fast* prices are increasing (or decreasing, although decreases are extremely rare). The table indicates that while prices are still rising in June, they are rising at a more moderate rate than previously. Thus, prices rose in June, but *inflation* fell in June. (Now that you know calculus, keep your ears open for how often broadcasters confuse these two statements and announce the wrong one!)

2.19 A motorist enters a turnpike, and his distance from the entrance is given by $s(t) = 15t^2 + 25t$, where t is measured in hours and s in miles.

a. At what time will he reach the speed limit of 55 miles per hour?

b. Suppose the motorist will be on the turnpike for 50 miles. Will he exceed the speed limit?

Solution:

a. $v = ds/dt = 30t + 25$. Set this equal to 55 and solve, obtaining t = 1 hour.

b. After 1 hour, his distance from the entrance is 40 miles. At this time his velocity is 55 MPH, and it continues to increase, since v is a linear function with positive slope. Hence, he *will* exceed the speed limit.

2.20 The position of a particle moving in a straight line is shown in Figure 2-29. Fill in the blanks in the following table by inserting a + or − in each space:

time	position	velocity	acceleration
t_1			
t_2			
t_3			

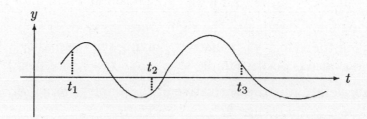

Figure 2-29: Position of a moving particle

Solution: The position can be determined by whether the curve lies above or below the t-axis, the velocity by whether the curve is increasing or decreasing, and the acceleration from the concavity of the curve. We obtain

time	position	velocity	acceleration
t_1	+	+	—
t_2	—	+	+
t_3	+	—	+

2.21　A particle is moving along a horizontal line according to the formula $s(t) = t^3 - 9t^2 + 24t - 10$,　$t > 0$, where $s(t)$ gives the position of the particle as a function of time.

a.　Find the position of the particle after 3 seconds.

b.　Find the instantaneous velocity of the particle as a function of time.

c.　At what time(s) does the particle momentarily come to rest?

d.　When is the particle moving to the right? To the left?

e.　Find the acceleration of the particle as a function of time.

Solution:

a.　$s(3) = 3^3 - 9 \cdot 3^2 + 24 \cdot 3 - 10 = 8$.

b.　The instantaneous velocity is the derivative of the position function. Thus, $v(t) = s'(t) = 3t^2 - 18t + 24$.

c.　The particle comes to rest when its velocity is 0. So we set $s'(t) = 0$ and solve the resulting equation. We obtain

$$3t^2 - 18t + 24 = 0$$
$$3(t - 2)(t - 4) = 0$$
$$t = 2, \, 4.$$

So the particle comes to a momentary stop at $t = 2$ and 4.

d. The particle is moving to the right when its velocity is positive, and to the left when the velocity is negative. Solving the inequality $3t^2 - 18t + 24 > 0$, we find that v is positive for $0 \leq t < 2$ and $t > 4$. v is negative for $2 < t < 4$.

e. The acceleration is the derivative of the velocity function: $v'(t) = 6t - 18$.

2.22 A mountain climber stumbles and as a consequence a small rock falls over the edge of a cliff which is 576 feet high. In t seconds the rock drops $16t^2$ feet.

a. How long does it take for the rock to reach the ground?

b. What is the *average* velocity of the rock while it is falling?

c. What is the *instantaneous* velocity of the rock at the moment it hits the ground?

Solution:

a. Solve the equation $16t^2 = 576$, obtaining $t = 6$ seconds.

b. The rock travels 576 feet in 6 seconds, so its average velocity is $576/6 = 96$ feet per second.

c. The instantaneous velocity is computed by taking the derivative of the distance function, $s(t) = 16t^2$. We find that $v(t) = 32t$, so that $v(6) = 192$ feet per second.

2.23 Figure 2-30 is a graph showing the *speeds* of two runners at various times during a race.

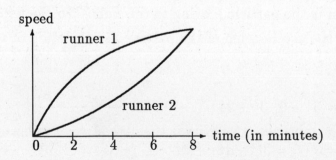

Figure 2-30: Speed of the runners

a. Which runner is going faster at time 6 minutes? At time 8 minutes?

b. Is runner 1 going faster at time 2 or 6 minutes?

    c.   In the time from 4 to 8 minutes, do the runners get farther apart or closer together?

    d.   If the race were to continue for another 10 minutes, which runner do you think will win?

Solution: The important thing to keep in mind while solving this problem is that the vertical axis represents speed and *not* position. This means that the slopes of the curves do not represent velocity, but rather acceleration.

    a.   Runner 1 is going faster throughout the time period from 0 to 8 minutes. At 8 minutes, the second runner's speed finally reaches that of the first runner.

    b.   Each runner's speed is increasing throughout the time period.

    c.   Since the first runner is always going faster, the distance separating them keeps increasing.

    d.   This is a hard one! In fact, we don't have enough information to answer the question, but we can analyze the situation. At the end of 8 minutes, runner 1 is well ahead of runner 2, since his speed has always been greater. At the 8 minute mark their speeds are equal. However, the second runner's acceleration is *much greater* than that of the first runner, and he finally begins to narrow the gap between them. Certain questions need to be answered before we can determine who will win the race. For example, will the second runner continue his now rapid pace while the first slows down? Is there enough time remaining for the second runner to make up the distance that he trails after 8 minutes?

2.24   One might think that the following is an alternate procedure for finding the slope of a curve at a point P. The usual definition, which we have adopted, arises from computing the slope of the secant line joining P with a second point Q, and then allowing Q to move along the curve toward P. This leads to the definition of the derivative

$$f'(x) = \lim_{h \to 0} \frac{f(x+h) - f(x)}{h}.$$

We could, instead, choose 2 points, Q_1 and Q_2, on either side of P, symmetrically placed (see Figure 2-31), compute the slope of the secant line joining Q_1 and Q_2, and then let Q_1 and Q_2 approach P *simultaneously*. Doing so would lead to the evaluation of the following limit:

$$f^*(x) = \lim_{h \to 0} \frac{f(x+h) - f(x-h)}{2h}$$

(We use the symbol f^* to distinguish this limit from f'.)

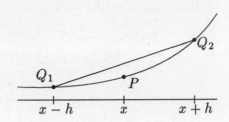

Figure 2-31: Symmetric approximation

a. Show that if $f'(x)$ exists, then so does $f^*(x)$. (This says that if f is differentiable in our original sense, then it is also "differentiable" in this new sense.)

b. Show, by example, that the two procedures are *not* equivalent; that is, find a function for which f^* exists at a point P, but f' does not.

Solution:

a. Suppose $f'(x)$ exists. Then

$$
\begin{aligned}
f^*(x) &= \lim_{h \to 0} \frac{f(x+h) - f(x-h)}{2h} \\
&= \lim_{h \to 0} \frac{f(x+h) - f(x) + f(x) - f(x-h)}{2h} \\
&= \lim_{h \to 0} \left[\frac{f(x+h) - f(x)}{2h} + \frac{f(x) - f(x-h)}{2h} \right] \\
&= \lim_{h \to 0} \frac{f(x+h) - f(x)}{2h} + \lim_{h \to 0} \frac{f(x) - f(x-h)}{2h} \\
&= \frac{f'(x)}{2} + \frac{f'(x)}{2} \\
&= f'(x).
\end{aligned}
$$

b. Let $f(x) = |x|$ and let P be the origin. We know that $f'(0)$ does not exist. However,

$$ f(x+h) = f(0+h) = f(h) = |h| $$

and

$$ f(x-h) = f(0-h) = f(-h) = |-h| = |h|, $$

so that

$$ f^*(0) = \lim_{h \to 0} \frac{f(0+h) - f(0-h)}{2h} = \lim_{h \to 0} \frac{|h| - |h|}{2h} = 0. $$

2.25 Let $f(x) = x|x|$.

    a. Explain why the product rule *cannot* be used to find $f'(0)$.

    b. Find $f'(0)$ directly from the definition of the derivative.

Solution:

    a. The product rule allows us to compute the derivative $f = gh$, *provided* both of the functions g and h are differentiable. In our case, however, $|x|$ fails to have a derivative at $x = 0$, so that we cannot apply the product rule.

    b. By definition,

$$f'(0) = \lim_{h \to 0} \frac{f(0 + h) - f(0)}{h}$$

$$= \lim_{h \to 0} \frac{h|h| - 0}{h}$$

$$= \lim_{h \to 0} |h| = 0.$$

2.26 Find an example of a function which has a derivative at $x = 0$, but which fails to have a second derivative at $x = 0$.

Solution: We saw in Example 2.3 that $|x|$ has no derivative at $x = 0$, while in the previous solved problem we showed that $f(x) = x|x|$ is differentiable there. We'll see that f can serve as the desired example. We first rewrite f as

$$f(x) = \begin{cases} x^2, & x \geq 0 \\ -x^2, & x < 0 \end{cases}$$

If $x \neq 0$ we can easily compute f'. For $x > 0$, we have $f'(x) = 2x$, while for $x < 0$, $f'(x) = -2x$. From the previous problem $f'(0) = 0$. Putting this all together, we obtain

$$f'(x) = \begin{cases} 2x, & x \geq 0 \\ -2x, & x < 0 \end{cases}$$

But this is nothing more than $2|x|$, which we know is not differentiable at $x = 0$.

2.27 **Always–sometimes–never:**

    a. The instantaneous velocity of a moving body is _____ less than the average velocity.

    b. If the derivative of a function is always 0, then the function is _____ a constant.

    c. The instantaneous velocity of a moving body _____ equals its average velocity at some time.

d.   The derivative of fg is —————— $f'g'$.

e.   The derivative of $(f(x))^n$ is —————— equal to $n\,(f(x))^{n-1}$.

Solution:

a.   sometimes

b.   always

c.   always (use the Mean Value Theorem)

d.   never

e.   sometimes (the statement is true if and only if $f(x) = x$).

Supplementary Problems

2.28   Find the slope of the curve $y = 2x^3$ at the point $P = (x_0, y_0)$.

2.29   Find the derivatives of the following functions:

a. $\sin 4x$

b. $(x^2 - 7x)/(x^4 + 2)$

c. $x \ln x$

d. $e^{2x} + x^2$

e. $\tan(x^2 + 1)$

f. $\sqrt{\sin x + 4}$

g. $(x - \frac{1}{x})^5$

2.30   For each of the following functions find the equation of the line tangent to the curve $y = f(x)$ at the point $(a,\ f(a))$:

a. $f(x) = x^3 - 2x$ at $(2,4)$.

b. $f(x) = \cos x$ at $(\frac{\pi}{2}, 0)$.

c. $f(x) = xe^x$ at $(-1, -\frac{1}{e})$.

2.31   For each of the following functions find intervals on which the function is increasing and intervals on which it is decreasing:

    a. $f(x) = 2x^3 - 3x^2 + 2$

    b. $f(x) = xe^{-x}$

2.32   Find the 29th and 30th derivatives of $x^{29} + x^{28} + x^{27} + \cdots + x^2 + x + 1$. (No computations necessary!)

2.33   The following problem concerns the variation in the number of hours of daylight as the seasons change.

    a.   Make a rough sketch of the graph of a function in which the horizontal axis represents months of the year (January, February, etc.) and the vertical axis represents the number of hours of daylight in your city at the corresponding time. (You can obtain these numbers from an almanac, or just use approximate figures based upon your experience.) Do this for a 2-year period.

    b.   At what time of the year are the days longest?

    c.   At what time of the year is the length of the day (that is, the number of hours of daylight) increasing at the *most rapid rate*?

2.34   Sketch graphs of functions on an interval $[a, b]$ which satisfy:

    a. $f(x) > 0$, $f'(x) > 0$, $f''(x) < 0$

    b. $f(x) < 0$, $f'(x) > 0$, $f''(x) > 0$.

2.35   Figure 2-32 is the graph of a function f.

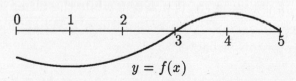

Figure 2-32:

    a.   Sketch the graph of f'.

    b.   Sketch the graph of a second function, g, which satisfies $g(0) = 0$, $g'(x) = f'(x)$ for all x in $[0,5]$.

2.36   Suppose a car is moving on a straight road, with $s(t)$ representing the distance traveled after t minutes. Describe in words the meaning of each of the following conditions:

a. $s'(t)$ is negative.

b. $s'(t)$ is constant.

c. $s'(t) = 0$ for $10 \leq t \leq 12$.

d. $s'(t)$ and $s''(t)$ are both positive.

2.37   Draw a graph of a function which describes the following trip taken by a motorist over a 3-hour period. The horizontal axis should be t for time in hours and the vertical one s for distance in miles.

- The motorist drives at a constant speed for $1\frac{3}{4}$ hours.

- He then stops for coffee for 15 minutes.

- When he resumes his trip he finds himself in a bit of a traffic jam. However, the traffic jam eases gradually, so that he is able to increase his speed continually over the next 15 minutes.

- He resumes the speed that he drove at in part 1, until he arrives at his destination 45 minutes later.

2.38   The total number (in thousands) of bacteria present in a culture after t hours of growth is given by $N(t) = 2t(t - 10)^2 + 50$.

a.   Find $N'(t)$.

b.   At what rate is the population of bacteria changing when

    (i) $t = 8$ hours?

    (ii) $t = 11$ hours?

c.   One of the answers in (b) is positive while the other is negative. (If you got different results, check your calculations!) What does this mean in terms of the population of bacteria?

2.39   Figure 2-33 shows the growth of bacteria in two different colonies.

a.   Which dish has *more bacteria* after 10 minutes?

b.   Which bacteria colony is *growing faster* after 10 minutes?

c.   If the two colonies continue to grow at the rates indicated, which of them will be larger at the end of 1 hour?

d.   What is the *average rate of growth* of the bacteria in dish 1 over the first 15 minutes? Of those in dish 2?

e.   From the graph, estimate the *instantaneous rate of growth* of the two colonies at time 15 minutes. (State your answer in bacteria per minute.)

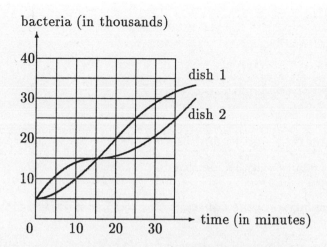

Figure 2-33: Growth of bacteria

2.40 Let $C(t)$ be the Consumer Price Index (CPI), where the time, t, is given in months. The CPI is used to measure inflation. Translate each of the following sentences into a statement about $C(t)$ and its derivatives. Draw a graph illustrating each case.

    a. Prices are still rising, but the inflation rate dropped last month.

    b. For the first time in many years, prices dropped last month.

    c. While inflation is modest now, we anticipate a much bigger rise in prices in the next few months.

2.41 Prove the Constant Multiple Rule: The derivative of cf is cf'.

Answers to Supplementary Problems

2.28 $6x_0^2$

2.29 **a.** $4\cos 4x$.

    b. $(-2x^5 + 21x^4 + 4x - 14)/(x^4 + 2)^2$.

    c. $\ln x + 1$.

    d. $2e^{2x} + 2x$.

    e. $\sec^2(x^2 + 1) \cdot 2x$.

    f. $\cos x/(2\sqrt{\sin x + 4})$.

g. $5(x - \frac{1}{x})^4(1 + \frac{1}{x^2})$.

2.30 **a.** $y = 10x - 16$.

b. $y = -x + \frac{\pi}{2}$.

c. $y = -1/e$.

2.31 **a.** Increasing on $(-\infty, 0)$ and $(1, \infty)$, decreasing on $(0, 1)$.

b. Increasing on $(-\infty, 1)$, decreasing on $(1, \infty)$.

2.32 The 29th derivative is the constant $29 \cdot 28 \cdot 27 \cdots 3 \cdot 2 \cdot 1$, so the 30th is 0.

2.33 **b.** The days are longest in June (in the Northern Hemisphere).

c. The length of the day is increasing most rapidly in March.

2.36 **a.** The car is moving backward.

b. The car is traveling at a constant speed.

c. The car has stopped for two minutes.

d. The car is moving forward at ever increasing speed.

2.38 **a.** $N'(t) = 6t^2 - 80t + 200$.

b. **(i)** $N'(8) = -56$.

(ii) $N'(11) = 46$.

c. The colony of bacteria is diminishing at $t = 8$, but growing at $t = 11$.

2.39 **a.** Dish 2.

b. Dish 1.

c. Dish 2.

d. 667 per minute for each colony.

e. Dish 1: 1000 per minute; dish 2: 0.

Answers to Exercises

2.1 (Page 33)

Function	Derivative
$x^4 + \cos x$	$4x^3 - \sin x$
$x^4 - \cos x$	$4x^3 + \sin x$
$x^4 \cos x$	$4x^3 \cos x - x^4 \sin x$
$x^4 / \cos x$	$(4x^3 \cos x + x^4 \sin x)/\cos^2 x$
$\cos x / x^4$	$(-x^4 \sin x - 4x^3 \cos x)/x^8$ $= (-x \sin x - 4 \cos x)/x^5$
$\cos(x^4)$	$-\sin(x^4) \cdot 4x^3$
$(\cos x)^4$	$-4 \cos^3 x \sin x$

2.2 (Page 35) $\dfrac{2 \cos x - x \sin x}{2\sqrt{\sin x + x \cos x}}$

Chapter 3

Applications of the Derivative

3.1 Newton's Method

What We Know: How to solve the linear equation $ax + b = 0$.

What We Want To Know: How to solve the equation $f(x) = 0$, for an *arbitrary* function, f.

How We Do It: We approximate the solution of $f(x) = 0$ with a sequence of solutions of certain linear equations.

3.1.1 Introduction

From our high school courses we have learned the importance of solving equations. A good deal of emphasis was placed on the solution of linear and quadratic equations, such as $2x - 5 = 0$ or $2x^2 - 7x + 4 = 0$, and techniques were developed for solving such equations. But in calculus more complicated equations arise, for example, in the solution of maximum-minimum problems. How can we solve equations such as $x^3 - 4x + 11 = 0$, or $x - \cos x = 0$? Is there some general way in which we can solve the equation $f(x) = 0$ for an arbitrary function, f? In this section we will discuss a very powerful method, discovered by Isaac Newton, which will enable us to solve many such equations. In addition, Newton's method will provide us with an important application of the derivative, as well as another example of the key role of our organizing principle: *Approximation — Refinement — Limit* $(\mathcal{A} - \mathcal{R} - \mathcal{L})$.

Before we begin, it is necessary to elaborate a bit on what we mean by a solution of $f(x) = 0$. From our high school work, we are used to obtaining exact solutions of both linear and quadratic equations. Thus, for example, the solution of $2x - 5 = 0$ is $x = 5/2$ or 2.5, while the solutions of $2x^2 - 7x + 4 = 0$ are $x = (7 \pm \sqrt{17})/4$. While the latter is an *expression* for the exact solutions of the quadratic equation, it is not of much practical value. An engineer, for example would need to turn that expression into a numerical (decimal) value, and in doing so would have to *approximate* $\sqrt{17}$ to whatever number of decimal places is desired. For the engineer, the approximation is generally more useful than the exact expression.

But even for the mathematician, who is usually happy with the exact expression, there are pitfalls. While every quadratic equation can be 'solved' as above, this is not true for equations in general. True, there are (much more complicated) formulas and procedures for solving third and fourth degree equations. However, it was shown in the early part of the 19th century that no formula comparable to the quadratic formula exists for solving equations of degree five and higher, such as $x^5 - 6x^3 + 7x^2 + 11x + 13 = 0$. This famous result precludes the possibility of obtaining exact solutions of such equations. Similarly, an equation such as $x - \cos x = 0$ cannot be solved exactly, and the *best* we can hope for is an approximation *to whatever degree of accuracy we desire*. Newton's method, which we are about to develop, provides just such a procedure.

Amazingly, Newton's method requires nothing more than knowledge of how to solve the simplest of equations: $ax + b = 0$. The steps will, by now, seem familiar. We construct an appropriate linear equation, whose solution provides an *approximation* to the desired root of $f(x) = 0$. Repeating this procedure produces a sequence of *refined* approximations, and the exact solution is obtained (at least, theoretically) by passing to the *limit* (provided it exists). We proceed now with the details.

3.1.2   The Method

The basic idea is best presented geometrically. Consider the graph of the function $y = f(x)$ (Figure 3-1), and suppose that $f(r) = 0$. We want to find r. Let x_0 be

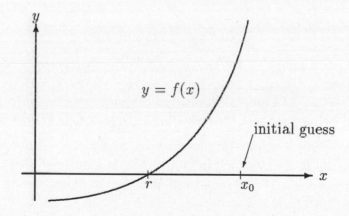

Figure 3-1: Initial approximation

an initial guess, or an *approximation* to r. If $f(x_0) = 0$ (in other words, our guess is an incredibly lucky one!), then we are done. But, more likely, $f(x_0) \neq 0$. Now construct the tangent, T_0, to $y = f(x)$ at the point $(x_0, f(x_0))$ (Figure 3-2). Let x_1 be the point at which T_0 crosses the x-axis. (It is here that we will have to solve a linear equation.) We consider x_1 a *refinement* of our initial approximation. To further refine this approximation, construct the tangent, T_1, to $y = f(x)$ at $(x_1, f(x_1))$, and let x_2 be the intersection of T_1 with the x-axis (Figure 3-3). Continuing our refinement

$\mathcal{A}$

$\mathcal{R}$

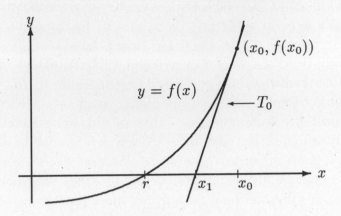

Figure 3-2: Refined approximation

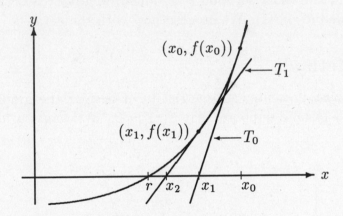

Figure 3-3: Further refinement

process in this way yields a sequence of points $x_0, x_1, x_2, \ldots$. To obtain the exact value of r, we pass to the *limit*. *Under certain conditions*, it can be shown that this sequence approaches r as the process is continued indefinitely. In other words,

$$r = \lim_{n \to \infty} x_n.$$

But how do we actually *compute* $x_1, x_2, x_3, \ldots$? To do this, we need the *equations* of the tangent lines, $T_0, T_1, T_2, \ldots$. Since for any n, T_n passes through the point $(x_n, f(x_n))$, its equation, from the point-slope formula, is

$$y - f(x_n) = f'(x_n)(x - x_n)$$

or

$$y = f'(x_n)x + (f(x_n) - x_n f'(x_n)).$$

To find the next element in the sequence, x_{n+1}, we set this equation equal to 0 and solve for x:

$$f'(x_n)x + (f(x_n) - x_n f'(x_n)) = 0$$

or

$$f'(x_n)x = x_n f'(x_n) - f(x_n)$$

or

$$x = x_n - \frac{f(x_n)}{f'(x_n)}$$

Hence,

$$x_{n+1} = x_n - \frac{f(x_n)}{f'(x_n)}. \tag{3.1}$$

We see that each term in the sequence is determined by the previous term, a process called *iteration*.

Perhaps the best known application of Newton's method is to the rapid calculation of square roots.

Example 3.1 *Use Newton's method to compute $\sqrt{A}$.*

Solution: To calculate $\sqrt{A}$, let $f(x) = x^2 - A$ and set $f(x) = 0$. (The solution is clearly $x = \sqrt{A}$.) The above equation (3.1), in this particular case, becomes

$$x_{n+1} = x_n - \frac{x_n^2 - A}{2x_n} \tag{3.2}$$

or

$$x_{n+1} = \frac{2x_n^2 - x_n^2 + A}{2x_n}$$

$$= \frac{x_n^2 + A}{2x_n}$$

or

$$x_{n+1} = \frac{x_n + A/x_n}{2} \tag{3.3}$$

As a special case, let $A = 2$, and let the initial approximation, x_0, be 1. Then

$$
\begin{array}{lll}
x_1 = & (1+2)/2 & = 1.500000000 \\
x_2 = & (1.5 + 2/1.5)/2 & = 1.416666667 \\
x_3 = & (1.416666667 + 2/1.416666667)/2 & = 1.414215686 \\
x_4 = & (1.414215686 + 2/1.414215686)/2 & = 1.414213562,
\end{array}
$$

which is the value of $\sqrt{2}$, correct to 9 decimal places. The speed with which we obtained this highly accurate approximation (just four iterations) is typical of Newton's method.

Example 3.2 *Use Newton's method to solve the trigonometric equation $x - \cos x = 0$.*

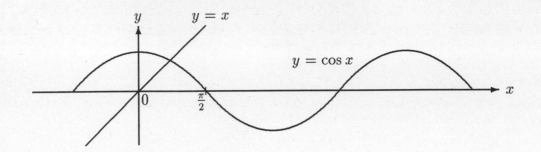

Figure 3-4: Graphs of $y = \cos x$ and $y = x$

Solution: From the graphs of $y = x$ and $y = \cos x$ (Figure 3-4), it appears that there is a root (solution) in the general vicinity of $x = 1$. So we let $x_0 = 1$ be the initial approximation. Since $f(x) = x - \cos x$, we have $f'(x) = 1 + \sin x$, so that the general iteration formula (3.1),

$$x_{n+1} = x_n - \frac{f(x_n)}{f'(x_n)}$$

becomes

$$x_{n+1} = x_n - \frac{x_n - \cos x_n}{1 + \sin x_n}$$

or

$$x_{n+1} = \frac{x_n \sin x_n + \cos x_n}{1 + \sin x_n}. \tag{3.4}$$

Four iterations of (3.4) yield

$$
\begin{aligned}
x_0 &= 1.0000000000 \\
x_1 &= 0.7503638678 \\
x_2 &= 0.7391128909 \\
x_3 &= 0.7390851334 \\
x_4 &= 0.7390851332,
\end{aligned}
$$

which is accurate to 10 decimal places.

Since the method is both easy to apply and very efficient, it would be nice if it always worked. Unfortunately, this is not the case. Indeed, a number of things can go wrong. The next example illustrates one of the pitfalls; others will be found among the Solved Problems.

Example 3.3 Return to Example 3.1, where we calculated $\sqrt{A}$, but this time choose $x_0 = 0$, rather than $x_0 = 1$. Now, $f'(0) = 0$ for $f(x) = x^2 - A$, so that the iteration breaks down at the very first step because $f'(x_n)$ appears in the *denominator* of (3.1).

Of course, in this case we can avoid this calamity by a more careful choice of x_0 (any *positive* value will do), but there are other situations in which the problem is inherent.

Newton's method is exceptionally powerful, but must be used with care. Precise conditions which guarantee the convergence of the method can be found in most textbooks.

3.2   Linear Approximation and Taylor Polynomials

Take out a *scientific* calculator (any one with a cosine button will do), turn it on, and put it in *radian* mode. (Calculators have *radian* and *degree* modes for trigonometric calculations—the instruction booklet will tell you how to place yours in radian mode.) Now compute some values, such as sin 1 or cos .5. Depending on how many decimal digits your calculator displays, you will obtain something like this:

$$\sin 1 \quad = .8414710 \text{ or } \sin 1 \quad = .8414709848$$
$$\cos .5 \quad = .8775826 \text{ or } \cos .5 \quad = .8775825619.$$

Now try $\sqrt{26}$, yielding either 5.0990195 or 5.099019514. *Where do these numbers come from?* Does the calculator have a *complete* table of sines, cosines and square roots built in? (If so, it must be a *very* extensive table, for if you ask it to calculate, say, sin 1.26898560157, it pops right back with .9547998172.) Before we begin investigating this problem in detail, please perform the following calculation:

$$1 - \frac{1}{6} + \frac{1}{120} - \frac{1}{5040} + \frac{1}{362,880}$$

The result is .8414710, which, surprisingly, agrees with the value of sin 1 to 7 decimal places. Similarly, the computation

$$1 - \frac{(.5)^2}{2} + \frac{(.5)^4}{24} - \frac{(.5)^6}{720} + \frac{(.5)^8}{40320}$$

yields .8775826, the value of cos .5 to 7 decimal places. But what do these sums have to do with sines and cosines? We'll soon see that they have a great deal to do with these functions. In fact, there are no *tables* built into our calculators, but rather, they employ *methods* for computing these functions. (The calculations we've just performed are indications of these methods.) In this section, we'll learn just how calculators got to be so smart (and fast).

When introducing the derivative, we mentioned that the tangent to a curve is the line which best approximates the curve near the point of tangency (page 13). We'll pursue this remark in the current section, and will see that this fundamental idea has far-reaching consequences.

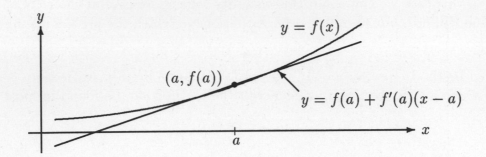

Figure 3-5: Tangent to $y = f(x)$

Recall from (2.17) in Section 2.3, that the equation of the tangent to the curve $y = f(x)$ at $(a, f(a))$ (Figure 3-5) is given by

$$y = f(a) + f'(a)(x - a). \tag{3.5}$$

We denote the function on the right-hand side of (3.5) by $P_1(x)$. Hence,

$$P_1(x) = f(a) + f'(a)(x - a). \tag{3.6}$$

We use the letter P because this function is a *polynomial*, and the subscript 1 because it is a polynomial of *degree* 1; that is, the highest power of x that occurs is 1. Later on, in a generalization of what we're doing here, we'll encounter polynomials of degrees $2, 3, 4, \ldots$, which we'll denote by $P_2(x), P_3(x), P_4(x), \ldots$. To be more exact, we should denote this polynomial by $P_1(x; a)$, since different values of a yield different polynomials. Although this notation is used by some authors, we prefer the simpler version, $P_1(x)$, unless there is a chance for confusion, in which case we, too, will write $P_1(x; a)$.

Approximation by polynomials is of great importance. As opposed to trigonometric, logarithmic and exponential, and other functions, numerical values of polynomials are especially easy to compute, requiring only the basic arithmetic operations. This advantage of polynomials over their less fortunate 'cousins' has become even more prominent with the development of high-speed computers.

We now wish to make precise what we meant earlier when we said that the tangent, T, to the curve $y = f(x)$ at $x = a$ 'approximates' the curve. But before looking at this problem in general terms, let's look at a special case.

Example 3.4 *Find the tangent line approximation to $f(x) = \sqrt{x}$ at the point $(25, 5)$, and use it to approximate $\sqrt{26}$ and $\sqrt{30}$ (Figure 3-6).*

Solution: Since $f(x) = \sqrt{x} = x^{1/2}$, we have $f'(x) = (1/2)x^{-1/2}$. Thus, $f(25) = 5$ and $f'(25) = 1/10 = .1$. From (3.6),

$$P_1(x) = f(25) + f'(25)(x - 25)$$

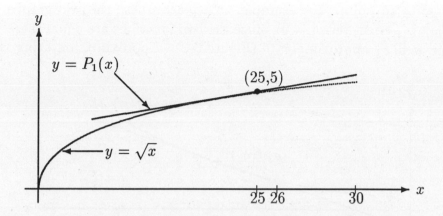

Figure 3-6: Tangent line approximation

or

$$P_1(x) = 5 + .1(x - 25). \tag{3.7}$$

Let us now use P_1 to approximate $\sqrt{26}$. From (3.7), we obtain $P_1(26) = 5 + .1(26 - 25) = 5.1$. We saw earlier that a calculator value of $\sqrt{26}$ is 5.0990195, so that the error in this approximation, $5.1 - 5.0990195 = .0009805$ is quite small.

At $x = 30$, however, the error is larger. From (3.7), we see that $P_1(30) = 5 + .1(30 - 25) = 5.5$, while the true value of $\sqrt{30}$ (to 7 decimal places) is 5.4772258, yielding an error of .0227744. So we see that the error, in general, depends upon x.

We measure how well one function, g, approximates a second, f, on an interval $[u, v]$ as follows: Let x be any point in the interval $[u, v]$, and consider $|f(x) - g(x)|$, the absolute value of the difference between these two functions at the point x. In other words, we are computing the vertical distance between the graphs of the two functions (Figure 3-7). g is a good approximation to f over the interval $[u, v]$ if the

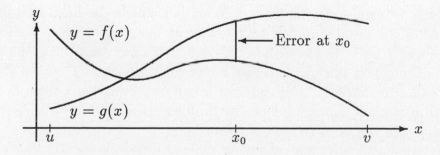

Figure 3-7: Error in approximating f by g

error, $|f(x) - g(x)|$, is 'small' for every x in $[u, v]$. In general, the approximation varies from point to point. For example, at a place where the two graphs touch, the approximation is perfect (error = 0), since the two functions are equal there. Let's now examine how well P_1 approximates f (Figure 3-8). Approximation using the tangent

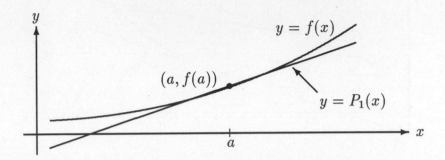

Figure 3-8: Approximation of f by P_1

line is also known as *linear approximation*, since P_1 is a *linear* function. (While we shall discuss this topic in some depth, additional details and examples can be found in your text.) At the point of contact, a, there is no error, since $P_1(a) = f(a)$. We see from Figure 3-8, however, that as x moves away from a, an error occurs; near a the error is small, but it grows rapidly for values of x remote from a. We want to get a handle on the size of the error, by finding an estimate of how large it can *possibly* be, a so-called *error bound*. We denote the error at the point x by $E_1(x)$; that is,

$$E_1(x) = f(x) - P_1(x). \tag{3.8}$$

What features determine the error, $E_1(x)$? Figure 3-8 made clear that the *distance* between x and a is important. Thus, it makes sense that the term $x - a$ should appear in the error estimate, but exactly how is not obvious. The dependence might appear in the form $(x - a)$ or $(x - a)^2$ or $\sqrt{x - a}$ or any other expression which yields 0 when $x = a$. And what else influences the error? Does it depend on f in any way? How? Consider the following two diagrams (Figures 3-9 and 3-10). In Figure 3-9 the error is quite large, even for values of x reasonably close to a, while in Figure 3-10 the error is moderate, even for values of x far from a. What is it about the function g in Figure 3-10 which makes the approximation by its tangent line, T_g, so much better than the approximation of f by T_f in Figure 3-9? It's due to the fact that the graph of $y = g(x)$ does not bend much, it is close to being a straight line, while f curves sharply. If g was, indeed, a linear function, then the error would be 0 everwhere, since T_g would then coincide with $y = g(x)$. Since g is not *exactly* linear, there is an error, but it is small in comparison with the error in approximating f.

Now what is the *calculus* way of saying that a function is 'nearly linear'? A linear function is one with *constant* slope; a *nearly linear* function is one whose slope is *nearly constant*. But what in the world do we mean by that?

Let's go a step further. Since a linear function, l, has constant slope, its derivative is a *constant function*. Thus, $l'(x) = C$, and *its* derivative, denoted by $l''(x)$, is 0, because

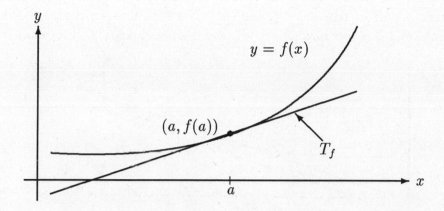

Figure 3-9: Poor approximation

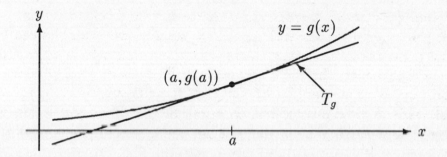

Figure 3-10: Good approximation

the derivative of a constant function is identically 0. Since the second derivative of a linear function is identically 0, by a 'nearly linear' function we mean one whose second derivative is *small*. Hence, we see that the second derivative of f also plays an important role in the formula for E_1.

After this long discussion, we finally pull these ideas together and state the result which contains the error estimate we are after. (See your text for the precise derivation.)

Theorem 3.1 *Suppose the second derivative of the function f exists. Let $E_1 = f - P_1$, where $P_1(x) = f(a) + f'(a)(x - a)$. Then, for each x, there exists a point c between a and x, such that*

$$E_1(x) = \frac{f''(c)(x - a)^2}{2} \tag{3.9}$$

(Figure 3-11).

Remark 3.1 Equation (3.9) is reminiscent of the Mean Value Theorem, which states that if f is differentiable on $[a, b]$, then, for each x in $[a, b]$, there exists a point c between

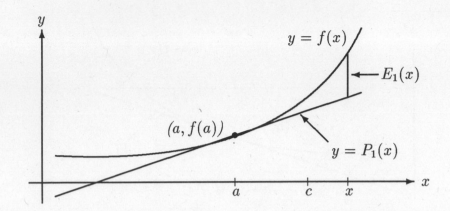

Figure 3-11: The error in linear approximation

a and x, such that

$$\frac{f(x) - f(a)}{x - a} = f'(c)$$

or

$$f(x) - f(a) = f'(c)(x - a). \qquad (3.10)$$

In fact, (3.9) can be viewed as a generalization of the Mean Value Theorem. (For those interested, a discussion of this remark can be found at the end of this section on Page 76.) Moreover, (3.10) and (3.9) are merely the first two steps in a much more extensive process, the development of which is the main purpose of this section.

Remark 3.2 Although we use the same letter c in both (3.9) and (3.10), these points are generally different.

Remark 3.3 In both (3.9) and (3.10), the point c depends on x. As x varies, so does c, often in complicated ways. In general, we cannot locate the point c precisely in either (3.9) or (3.10). However, we do know that it lies between a and x, and we will now see that even this limited knowledge often suffices to enable us to utilize these expressions in a practical way.

For example, let $f(x) = \sin x$, $a = 0$ (Figure 3-12). Then $f(0) = \sin 0 = 0$, while $f'(x) = \cos x$, so that $f'(0) = \cos 0 = 1$. Hence, $P_1(x) = f(0) + f'(0)x = x$, so that $y = x$ is the equation of the required tangent line. What about $E_1(x)$? We first compute $f''(x) = -\sin x$. By (3.9),

$$E_1(x) = \frac{f''(c)(x - a)^2}{2},$$

for some c between 0 and x. Hence,

$$E_1(x) = \frac{-\sin c \cdot x^2}{2}. \qquad (3.11)$$

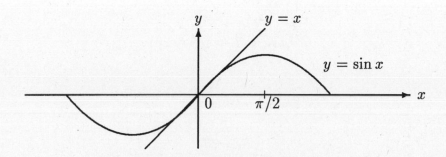

Figure 3-12: Graph of $f(x) = \sin x$

But we don't know c, so what good is (3.11)? Well, we can use it to get an estimate of $E_1(x)$, as follows. Regardless of the location of c, $|-\sin c| \leq 1$. Hence,

$$|E_1(x)| \leq \frac{x^2}{2}. \qquad (3.12)$$

(3.12) tells us that we can *control* the error by keeping x *sufficiently close* to 0. For example, if we want to use $P_1(x) = x$ as an approximation to $\sin x$, say in the interval $[-\pi/6, \pi/6]$, then

$$|E_1(x)| \leq \frac{(\pi/6)^2}{2} = \frac{\pi^2}{72} < .14.$$

(**Note:** Recall that all calculations of the trigonometric functions in calculus are done in radians) On the interval $[-\pi/12, \pi/12]$, the error is much smaller, since it cannot exceed

$$|E_1(x)| \leq \frac{(\pi/12)^2}{2} = \frac{\pi^2}{288} < .035.$$

Finally, on $[-\pi/36, \pi/36]$, we have

$$|E_1(x)| \leq \frac{(\pi/36)^2}{2} = \frac{\pi^2}{2592} < .004.$$

We see that P_1 is, at best, a fair approximation to $\sin x$ for values of x *near* 0 ($\pi/36$ is equivalent to just 5 degrees), but a poor one as x moves away from 0.

Let's return now to Example 3.2. We'll see that for the function $\sqrt{x}$, P_1 does a much better job. We obtain the error formula from (3.9):

$$E_1(x) = \frac{f''(c)(x-25)^2}{2}$$

or

$$E_1(x) = -\frac{(x-25)^2}{8c^{3/2}}. \qquad (3.13)$$

Let's now see how well P_1 approximates $\sqrt{x}$ for various values of x. This method gives us a simple way of computing square roots, but it will be of value *only if the errors involved are small*.

We start with $x = 26$. From (3.7),

$$P_1(x) = 5 + .1(x - 25)$$

so that

$$P_1(26) = 5 + .1(26 - 25) = 5.1,$$

while from (3.13)

$$E_1(x) = -\frac{(26 - 25)^2}{8c^{3/2}} = -\frac{1}{8c^{3/2}},$$

for some c between 25 and 26. We now encounter a new problem: In our analysis of the tangent line approximation of $\sin x$ above, the term involving the unknown point c was simply $\sin c$. Since the estimate $|\sin c| \leq 1$ is valid *everywhere*, this term could be eliminated from consideration. Here, however, we don't know where c is, and the expression for the error depends upon c. So what can we do? Well, first, let's retract the statement we just made. We *do* know something about c, quite a bit, in fact: c lies between 25 and 26. Now, we can't expect to compute the *exact* value of $E_1(26)$. (If we *could*, then we could find the *exact* value of $\sqrt{26}$, since $\sqrt{26} = P_1(26) + E_1(26)$, and we know $P_1(26)$.) The best we can hope for is a *bound* on how large $|E_1(26)|$ can *possibly be*. Now,

$$|E_1(26)| = \frac{1}{8c^{3/2}}, \quad 25 < c < 26,$$

and for c in this interval the function $1/(8c^{3/2})$ is largest when c is as small as possible. Hence,

$$\begin{aligned}
\frac{1}{8c^{3/2}} &< \frac{1}{8 \cdot 25^{3/2}} \text{ for } 25 < c < 26 \\
&= \frac{1}{8 \cdot 125} \\
&= \frac{1}{1000} \\
&= .001.
\end{aligned}$$

Thus, $|E_1(26)|$ is sure to be less than .001, so that the estimate of $\sqrt{26}$ given by $P_1(26)$ is accurate to within .001. In other words, the true value of $\sqrt{26}$ is within .001 of 5.1.

In this example there are independent ways of computing square roots, so that we can verify that the error bound is accurate. Now $\sqrt{26} = 5.0990195$, so that

$$5.1 - \sqrt{26} < .00099,$$

which, indeed, is less than .001 (though not by much!).

Let's stick with this example and compute several other square roots. Again, from (3.7), we obtain

$$P_1(27) = 5 + 1/5 = 5.2$$

and

$$P_1(30) = 5 + 1/2 = 5.5,$$

while from (3.13),

$$E_1(27) < 2^2/1000 = .004$$

and

$$E_1(30) < 5^2/1000 = .025.$$

Since $\sqrt{27} = 5.1961524$ and $\sqrt{30} = 5.4772256$, the actual errors in the approximation by $P_1(x)$ are

$$5.2 - \sqrt{27} < .00385,$$

and

$$5.5 - \sqrt{30} < .02278,$$

both of which are less than the guaranteed error bounds of .004 and .025, respectively.

Let's summarize some of the important conclusions we can draw at this point:

- The approximation technique we've developed *sometimes* yields acceptable estimates. If the value of x is far from the point of tangency, a, then the approximation deteriorates and may become useless in practical situations.

- The approximation may be much better for one function than for another, depending primarily upon their second derivatives. Thus, for $f(x) = \sqrt{x}$, we obtained reasonably good approximations, while for $f(x) = \sin x$ the approximations were mediocre, at best.

So we now ask the natural question:

How can we refine the approximation?

To answer this question, let's go back and look at $P_1(x)$, the tangent line approximation to $y = f(x)$ at the point $x = a$. It has the following properties:

1. It is a polynomial of degree 1;

2. $P_1(a) = f(a)$; that is, P_1 has the same *value* as f at a. Hence their graphs have the same *location* at a.

3. $P_1'(a) = f'(a)$; that is, the *derivative* of P_1 has the same value as f' at a. Hence their graphs have the same *direction* at a.

4. The error in the approximation, E_1, depends upon f'', the second derivative of f. The larger f'' is on the interval $[u, v]$, the larger the potential error.

Let's analyze the last property for a moment. The problem with P_1 is that, as a linear function, its graph *cannot bend*. On the other hand, if f'' is large, then the graph of $y = f(x)$ bends *a lot*. So if we are going to reduce the error we'll have to come up with an approximating function which can mirror this bend. Such a function clearly cannot be a first-degree polynomial, whose graph is a straight line.

But how about a second-degree polynomial, a quadratic function?

Looking at the first three of the properties of P_1 listed above and reasoning by analogy, let's see if we can find a polynomial, P_2, satisfying:

1. P_2 is a polynomial of degree *two*, a quadratic;

2. $P_2(a) = f(a)$;

3. $P_2'(a) = f'(a)$.

Finding such a polynomial is no problem; in fact, there are an infinite number of them. Just add on the term $K(x - a)^2$ to P_1, where K is any constant. In other words, let

$$P_2(x) = P_1(x) + K(x - a)^2 = f(a) + f'(a)(x - a) + K(x - a)^2 \qquad (3.14)$$

If we substitute $x = a$ into (3.14) we find that

$$P_2(a) = f(a) + f'(a)(a - a) + K(a - a)^2 = f(a).$$

Also, if we differentiate both sides of (3.14), we obtain

$$P_2'(x) = f'(a) + 2K(x - a). \qquad (3.15)$$

Substituting $x = a$ in (3.15) yields

$$P_2'(a) = f'(a) + 2K(a - a) = f'(a).$$

Hence, no matter what K is, $P_2(a) = f(a)$ and $P_2'(a) = f'(a)$, so that P_2 and f share the same *values and derivatives* at $x = a$. But we haven't exploited the freedom given us in the choice of K. In fact, we haven't chosen K at all yet! Suppose we now try to pick K so that P_2 also satisfies $P_2''(a) = f''(a)$. If we can do this, then the graph of $y = P_2(x)$ will have the following features in common with that of $y = f(x)$ *at the point $x = a$*:

1. The same *location*, since $P_2(a) = f(a)$;

2. The same *direction*, since $P_2'(a) = f'(a)$;

3. The same *bending*, since $P_2''(a) = f''(a)$.

Moreover, it's easy to find K. Differentiate P_2 twice:

$$
\begin{aligned}
P_2(x) &= f(a) + f'(a)(x - a) + K(x - a)^2 \\
P_2'(x) &= f'(a) + 2K(x - a) \\
P_2''(x) &= 2K,
\end{aligned}
$$

so that P_2'' is a constant, $2K$. Since we require that $P_2''(a) = f''(a)$, we have $2K = f''(a)$ or $K = f''(a)/2$, so that

$$P_2(x) = f(a) + f'(a)(x - a) + \frac{f''(a)(x - a)^2}{2} \qquad (3.16)$$

is the desired polynomial.

Example 3.5 *Find P_2 for $f(x) = \cos x$ at $a = 0$.*

Solution: We first compute $f'(x) = -\sin x$ and $f''(x) = -\cos x$. Since $a = 0$, we have $f(0) = \cos 0 = 1$, $f'(0) = -\sin 0 = 0$ and $f''(0) = -\cos 0 = -1$. Hence, from (3.16),

$$P_2(x) = 1 - \frac{x^2}{2}$$

(Figure 3-13).

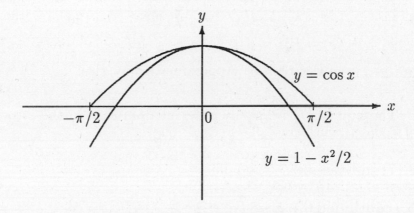

Figure 3-13: Approximation of $\cos x$ by P_2

Example 3.6 *Find P_2 for $f(x) = \sqrt{x}$ at $a = 25$.*

Solution: Since $f(x) = x^{1/2}$, we have $f'(x) = (1/2)x^{-1/2}$ and $f''(x) = (-1/4)x^{-3/2}$. Hence, $f(25) = 5$, $f'(25) = 1/10 = .1$ and $f''(25) = -1/500 = -.002$. Thus, from (3.16),

$$P_2(x) = 5 + .1(x - 25) - .001(x - 25)^2$$

(Figure 3-14).

We now turn to the error estimate. We write $E_2 = f - P_2$, and ask how we should expect the error to behave. Recall that $E_1(x) = f''(c)(x - a)^2/2$, for some c between a and x. Here, too, the distance between x and a will certainly influence the error and we'll see the exact dependence shortly. But what about the part of the error formula that depends on f? Reasoning as we did earlier, we see that the approximation will be exact if f, itself, is a second degree polynomial. Since the third derivative, f''', of such a polynomial is 0, it seems plausible that this term will enter the formula. The precise equation is given by

$$E_2(x) = \frac{f'''(c)(x - a)^3}{3!}, \tag{3.17}$$

for some c between a and x, where the term $3!$, called '3 factorial,' is equal to $3 \cdot 2 \cdot 1 = 6$. (Factorials will appear frequently in this section in subsequent error formulas.)

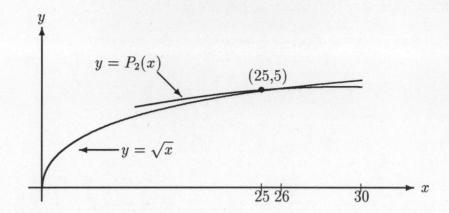

Figure 3-14: Approximation by P_2

Remark 3.4 The c in (3.17) is usually different from that in the error estimate for P_1 given by (3.8).

Let's apply (3.17) to Examples 3.5 and 3.6.

Example 3.5 (continued) *Analyze the error E_2 for $f(x) = \cos x$ at $a = 0$.*

Solution: Here, $f'''(x) = \sin x$, so that from (3.17)

$$E_2(x) = \frac{\sin c (x - a)^3}{6}.$$

Since $|\sin c| \leq 1$ for all c, we have

$$|E_2(x)| \leq \frac{x^3}{6}. \tag{3.18}$$

Thus, for example, $E_2(\pi/6) \leq (\pi/6)^3/6 < .024$. Now $P_2(\pi/6) = 1 - (\pi/6)^2/2 = .8629$, and $\pi/6$ (30 degrees) is one of the values for which we have exact knowledge of the trigonometric functions. In fact, $\cos \pi/6 = \sqrt{3}/2 = .8660$, so that $\cos \pi/6 - P_2(\pi/6) = .0031$, well within the guaranteed error bound of .024.

Let's now evaluate $P_2(x)$ at a point at which we don't already know $\cos x$, say $x = .1$.

$$P_2(.1) = 1 - \frac{(.1)^2}{2} = 1 - \frac{.01}{2} = .995$$

with

$$|E_2(.1)| \leq \frac{(.1)^3}{6} = \frac{.001}{6} < .00017,$$

which is a *bound on the possible error*. Thus $\cos(.1)$ is within .00017 of .995.

Example 3.6 (continued) *Estimate $E_2(x)$ for $f(x) = \sqrt{x}$.*

Solution: For $f(x) = \sqrt{x}$ we have $f'''(x) = (3/8)x^{-5/2}$. Hence, (3.17) becomes

$$E_2(x) = \frac{(3/8)(x-25)^3}{6c^{5/2}}$$

or

$$E_2(x) = \frac{(x-25)^3}{16c^{5/2}}, \tag{3.19}$$

with c between 25 and x. As we saw earlier when we estimated E_1, the largest value of the right-hand side of (3.19) occurs when c is as small as possible, which for $x > 25$ is $c = 25$. Hence,

$$|E_2(x)| \le \frac{(x-25)^3}{16 \cdot 25^{5/2}} = \frac{(x-25)^3}{50000}$$

for $x > 25$.

For $x < 25$, the *smallest* possible value of c (which induces the *largest* possible error) is $c = x$. Thus, for example, if $x = 24$, then the error bound is

$$|E_2(24)| < \frac{|24-25|^3}{16 \cdot 24^{5/2}} < .000023,$$

while, for $x = 20$, the error bound is

$$|E_2(20)| < \frac{|20-25|^3}{16 \cdot 20^{5/2}} < .0044.$$

Encouraged by the success of our work with P_2, which was based on matching the second derivative of f at a, we now proceed. Isn't it plausible that matching additional derivatives will lead to even more *refined* approximations? (**Note:** While the first derivative has geometrical significance—direction, and the second derivative does too—bending, or *concavity*, as it is usually called, the third and higher derivatives have no obvious geometrical meaning. So the idea of matching additional derivatives is not based on direct geometrical reasoning, but rather on our initial success in passing from P_1 to P_2.) The next step in the procedure is to obtain P_3, which satisfies one new condition, $P_3'''(a) = f'''(a)$. An explicit form for this polynomial is

$$P_3(x) = f(a) + f'(a)(x-a) + \frac{f''(a)(x-a)^2}{2!} + \frac{f'''(a)(x-a)^3}{3!} \tag{3.20}$$

and the corresponding error formula turns out to be

$$E_3(x) = \frac{f''''(c)(x-a)^4}{4!}, \tag{3.21}$$

for some c between a and x.

Rather than illustrate (3.20) and (3.21) with examples, we move directly to the general case, in which we match the value of f and its first n derivatives at a. We obtain

$$P_n(x) = f(a) + f'(a)(x - a) + \frac{f''(a)(x-a)^2}{2!} + \cdots + \frac{f^{(n)}(a)(x-a)^n}{n!}, \qquad (3.22)$$

where $f^{(n)}(a)$ is the nth derivative of f at a. P_n is known as the nth *Taylor polynomial* of f at the base point, a.

The error formula for $E_n = f - P_n$ takes on a by now familiar form:

$$E_n(x) = \frac{f^{(n+1)}(c)(x-a)^{n+1}}{(n+1)!}, \qquad (3.23)$$

for some c between a and x.

Example 3.7   *Find P_3 and P_4 for $f(x) = \sqrt{x}, a = 25$.*

Solution: We have

$$
\begin{array}{llll}
f(x) & = x^{1/2}, & f(25) & = 5, \\
f'(x) & = x^{-1/2}/2, & f'(25) & = .1, \\
f''(x) & = -x^{-3/2}/4, & f''(25) & = -.002, \\
f'''(x) & = 3x^{-5/2}/8, & f'''(25) & = .00012, \\
f''''(x) & = -15x^{-7/2}/16, & f''''(25) & = -.000012.
\end{array}
$$

We obtain $P_3(x)$ and $P_4(x)$ from (3.22):

$$P_3(x) = 5 + .1(x - 25) - \frac{.002(x - 25)^2}{2!} + \frac{.00012(x - 25)^3}{3!}$$

or

$$P_3(x) = 5 + .1(x - 25) - .001(x - 25)^2 + .00002(x - 25)^3$$

and

$$P_4(x) = P_3(x) + \frac{f''''(25)(x - 25)^4}{4!}$$

or

$$P_4(x) = 5 + .1(x - 25) - .001(x - 25)^2 + .00002(x - 25)^3 - .0000005(x - 25)^4.$$

Computing $P_3(x)$ and $P_4(x)$ at $x = 30$ and 35, we obtain

$$
\begin{array}{ll}
P_3(30) = 5.47750, & \text{error} \leq .00027 \\
P_3(35) = 5.92000, & \text{error} \leq .00390 \\
P_4(30) = 5.47719, & \text{error} \leq .00004 \\
P_4(35) = 5.91500, & \text{error} \leq .00108
\end{array}
$$

Remark 3.5 The polynomial P_n given by (3.22) can be produced only if the function f has at least n derivatives at a, and the error formula (3.23) assumes that f actually has at least $n + 1$ derivatives there. For most of the elementary functions that we encounter in calculus, such as trigonometric, exponential and logarithmic functions, this does not present a problem.

Does the approximation *improve* as n gets larger? In other words, is the procedure *actually refined* when we increase the degree of the Taylor polynomial? To answer this question, we now analyze the three components of the error formula (3.23), especially its dependence on n, the degree of the polynomial.

- The term $(n + 1)!$ in the denominator is independent of both x and f, and it is a term that is always *guaranteed* to help make the error small. Indeed, since factorials grow very rapidly (for example, $10!$ is more than 3 million, and $20!$ is more than a *billion billion*—several hundred thousand times the national debt), $1/(n + 1)!$ is exceedingly small for large values of n.

- The term $(x - a)^{n+1}$, on the other hand, arising from the distance between x and the base point a, will be *helpful* if $|x - a| < 1$, since then $|x - a|^{n+1} \to 0$ as n becomes large, but *harmful* if $|x - a| > 1$, since $|x - a|^{n+1} \to \infty$ with increasing n in this case. So for points within 1 of a, this term is a plus. However, even if $|x - a| > 1$, it can be shown that the term $|x - a|^{n+1}$ is overwhelmed by the factorial. That is, as we've seen above, the factorial grows so fast that $|x - a|^{n+1}/(n + 1)! \to 0$, *no matter how large $|x - a|$ is.*

- Finally, we turn our attention to $f^{(n+1)}(c)$, the portion of the error formula which depends on the function we are approximating. Here, nothing general can be said. For some functions, such as $\sin x$ or $\cos x$, the derivatives of all orders are *bounded* (for these particular functions, by 1). In such a case, the approximations get better and better as n increases, for any x. (Even here, though, the approximations are generally superior for values of x close to a than for those far away from the base point.) For some other functions, however, the derivatives grow rapidly and, as a result, the approximations may improve with increasing n only within a limited interval surrounding a or, in extreme cases, may fail to improve at all. In other words, the refinement process can fail *completely* for certain peculiar functions, while for some others, its usefulness may be limited to some interval surrounding a. This is analogous to other examples of the failure of the refinement stage that occur: In Chapter 2, where we showed that the function $|x|$ has no derivative at $x = 0$, and in Solved Problem 3.4 where we'll see that Newton's method fails. A fuller discussion of these issues will take place in Chapter 7 on Infinite Series, at which time we will develop the tools necessary to analyze these questions. We are thus leaving for that chapter the *limit* of the Taylor procedure, having gone through the first two stages of *approximation* and *refinement* in the current section.

Optional: At the beginning of this section we said that the Mean Value Theorem fits into the Taylor process in a natural way. We now justify this assertion.

We've shown in this section how to produce a sequence of polynomials $P_1, P_2, P_3, \ldots,$ of degrees $1, 2, 3, \ldots,$ respectively, which match certain features of a function, f. Specifically, P_1 satisfies $P_1(a) = f(a)$ and $P_1'(a) = f'(a)$; P_2 satisfies an *additional* requirement, $P_2''(a) = f''(a)$, and so forth. But let's take a step *backward* from P_1, and consider a polynomial which satisfies just *one* condition, namely, that it is equal to f at $x = a$. We'll call this still mysterious polynomial P_0, and since we have moved backward from P_1, its degree should be *zero*. That is, P_0 is a *constant* polynomial, and the value of that constant is $f(a)$. Hence, $P_0(x) = f(a)$ (Figure 3-15).

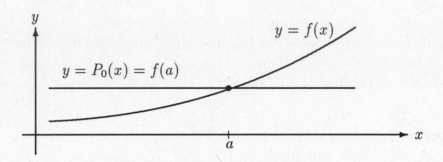

Figure 3-15: Approximation by P_0

Now, how well does P_0 approximate f? Does the expression for the error $E_0(x) = f(x) - P_0(x)$ fit into the general error formula

$$E_n(x) = \frac{f^{(n+1)}(c)(x-a)^{n+1}}{(n+1)!}?$$

When $n = 0$ this should read

$$E_0(x) = f'(c)(x-a),$$

for some c between a and x. Hence,

$$f(x) - P_0(x) = f'(c)(x-a)$$

or

$$f(x) - f(a) = f'(c)(x-a) \quad (\text{since } P_0(x) = f(a))$$

or

$$\frac{f(x) - f(a)}{x-a} = f'(c),$$

for some c between a and x. *But this is just the statement of the Mean Value Theorem!*

Solved Problems

NEWTON'S METHOD

3.1 **a.** Derive Newton's iteration for finding cube roots.

b. Apply the method to approximate $\sqrt[3]{5}$.

c. Apply the method to approximate $\sqrt[3]{30}$.

Solution:

a. Say we want to compute $\sqrt[3]{a}$. Let $f(x) = x^3 - a$, so that $f'(x) = 3x^2$. Then

$$x_{n+1} = x_n - \frac{f(x_n)}{f'(x_n)}$$

$$= x_n - \frac{x_n^3 - a}{3x_n^2}$$

$$= \frac{2x_n^3 + a}{3x_n^2}.$$

b. Clearly $1 < \sqrt[3]{5} < 2$, so $x_0 = 1$ seems like a reasonable starting point. Then

$$x_1 = \frac{2+5}{3} = \frac{7}{3} = 2.333333333$$

$$x_2 = \frac{2\left(\frac{7}{3}\right)^3 + 5}{3\left(\frac{7}{3}\right)^2} = \frac{821}{441} = 1.861678004$$

$$x_3 = 1.722001880$$
$$x_4 = 1.710059737$$
$$x_5 = 1.709975951$$
$$x_6 = 1.709975947,$$

which is correct to 9 decimal places.

If we start with $x_0 = 2$, then

$$x_1 = 1.75$$
$$x_2 = 1.710884354$$
$$x_3 = 1.709976429$$
$$x_4 = 1.709975947,$$

so that we obtain 9 place accuracy even more rapidly.

c.   A natural starting point is $x_0 = 3$ which is equal to $\sqrt[3]{27}$. From

$$x_{n+1} = \frac{2x_n^3 + 30}{3x_n^2}$$

we obtain

$$\begin{aligned}
x_1 &= 3.111111111 \\
x_2 &= 3.107237339 \\
x_3 &= 3.107232506,
\end{aligned}$$

correct to 9 decimal places.

An important application of the derivative (although one which is not covered in this book) is solving maximum-minimum problems. We usually accomplish this by finding the zeros of the *derivative* of the function to be maximized or minimized. Details of these procedures may be found in your text. Since, in some cases, finding these zeros is itself a difficult problem, Newton's method can be of assistance to us, so we apply the method to the following problem.

3.2   Find the zeros of the derivative of

$$g(x) = x^4 + 4x^3 - 4x^2 - 16x + 8.$$

Solution: Differentiating and equating to 0, we obtain

$$g'(x) = 4x^3 + 12x^2 - 8x - 16 = 4(x^3 + 3x^2 - 2x - 4) = 0.$$

So we are looking for the zeros of the function $x^3 + 3x^2 - 2x - 4$, which we call f. To apply Newton's method, we first calculate $f'(x) = 3x^2 + 6x - 2$. Hence, (3.1) becomes

$$x_{n+1} = x_n - \frac{x_n^3 + 3x_n^2 - 2x_n - 4}{3x_n^2 + 6x_n - 2}$$

or

$$x_{n+1} = \frac{2x_n^3 + 3x_n^2 + 4}{3x_n^2 + 6x_n - 2}.$$

As our initial guess, we take $x_0 = 0$, obtaining

$$x_1 = \frac{4}{-2} = -2,$$

and

$$x_2 = \frac{-16 + 12 + 4}{12 - 12 - 2} = \frac{0}{-2} = 0,$$

so that we are back where we started! Henceforth, the terms of the iteration will just bounce back and forth between -2 and 0, and the sequence does not approach a root of f.

Well, if $x_0 = 0$ doesn't work, what should we try? Perhaps we should start by analyzing the function f. Notice that $f(0) < 0$, while $f(2) > 0$, so that f has a zero on the interval $(0, 2)$. Similarly, $f(-2) > 0$, but $f(-4) < 0$. Thus, f has two more zeros, one in $(-4, -2)$ and another in $(-2, 0)$. So we'll choose $x_0 = -3$, -1 and 1 as our starting points in our attempt to locate the three zeros. We obtain the following results:

x_0	-3	-1	1
x_1	-3.28571429	-1	1.28571429
x_2	-3.23763999	-1	1.23763999
x_3	-3.23606963	-1	1.23606963
x_4	-3.23606798	-1	1.23606798

Note that we were lucky in our choice of $x_0 = -1$, since we hit one of the zeros right on the head!

3.3   Solve the equation $f(x) = x^3 - 5x = 0$, using Newton's method (Figure 3-16).

Solution: This problem can be solved easily by factoring, since $x^3 - 5x =$

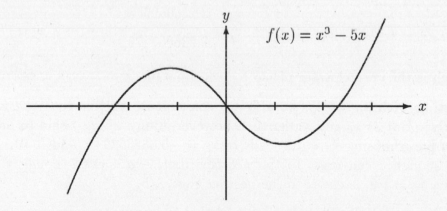

Figure 3-16: Graph of $f(x) = x^3 - 5x$

$x(x^2 - 5)$, yielding $x = 0$ and $x = \pm\sqrt{5}$ as solutions. But two important points will arise from the application of Newton's method to this equation:

- A harmless looking starting value, x_0, may fail to lead to a solution, even though nearby starting values do.

- We cannot easily predict which root the iteration will lead to.

Since $f(x) = x^3 - 5x$, $f'(x) = 3x^2 - 5$, so that (3.1) becomes

$$x_{n+1} = x_n - \frac{x_n^3 - 5x}{3x_n^2 - 5}$$

or

$$x_{n+1} = \frac{2x_n^3}{3x_n^2 - 5} \qquad (3.24)$$

Suppose we begin with $x_0 = 1$. Then (3.24) yields

$$\begin{aligned} x_1 &= & 2/(-2) &= & -1 \\ x_2 &= & -2/(-2) &= & 1, \end{aligned}$$

so that we are back where we started. The iteration will subsequently bounce back and forth between -1 and 1 like a ping-pong ball.

Notice that, unlike $x_0 = 0$ in Example 3.3, here $x_0 = 1$ does not appear to be a problematic point at all. There seems to be no way that we could have predicted the peculiar behavior of the iteration.

Now let's try some other starting values. For example, if we choose $x_0 = 2$, then, from (3.24), we have

$$\begin{aligned} x_1 &= & 2.285714286 \\ x_2 &= & 2.237639989 \\ x_3 &= & 2.236069633 \\ x_4 &= & 2.236067978, \end{aligned}$$

which is the correct value of $\sqrt{5}$ to 9 decimal places.

Starting with $x_0 = 1.5$, we also get convergence to this root (although more slowly). But as x_0 gets closer to 1, however, funny things begin to happen. For example, choosing $x_0 = 1.2$ leads to $x_1 = -5.082352941$, and it turns out that this sequence converges to the *negative* root, $-\sqrt{5}$, even though the starting value, $x_0 = 1.2$, is closer to the *positive* root, $\sqrt{5}$.

Now try $x_0 = 1/2$. This time, we find that x_n converges rapidly to 0. A picture of the convergence pattern is as follows:

- If $x_0 < -\sqrt{5/3}$, then x_n approaches $-\sqrt{5}$;
- If $-1 < x_0 < 1$, then x_n approaches 0;
- If $x_0 > \sqrt{5/3}$, then x_n approaches $\sqrt{5}$;
- If $x_0 = \pm\sqrt{5/3}$, then x_1 does not exist (the denominator in (3.24) is 0);
- If $x_0 = \pm 1$, the iteration oscillates between -1 and 1;

• If $-\sqrt{5/3} < x_0 < -1$ or $1 < x_0 < \sqrt{5/3}$, then the behavior becomes chaotic. For example, if $x_0 = 1.001$, then x_n approaches $\sqrt{5}$; but if $x_0 = 1.002$, then x_n approaches $-\sqrt{5}$. Similarly, if $x_0 = 1.00003$, then x_n approaches $\sqrt{5}$; for $x_0 = 1.00004$, however, x_n approaches $-\sqrt{5}$.

We see that Newton's method provides one of the easiest introductions to the phenomenon of *chaos*.

3.4 Solve the equation $f(x) = 0$ for the function $f(x) = x^{1/3}$, using Newton's method (Figure 3-17).

Solution: (While it's foolish to use the method here, since f is obviously 0

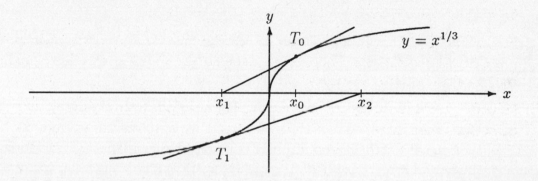

Figure 3-17: Newton's method fails

at $x = 0$, this example, nevertheless, will be instructive.) (3.1) becomes, quite simply, $x_{n+1} = -2x_n$. Suppose we choose $x_0 = 1/2$. We then obtain

$$
\begin{aligned}
x_1 &= -1 \\
x_2 &= 2 \\
x_3 &= -4 \\
x_4 &= 8,
\end{aligned}
$$

and so forth, so that the sequence clearly diverges. In fact, *no* choice of x_0 will lead to convergence, other than $x_0 = 0$.

3.5 Suppose $f(x) = 2x - 10$, and let $x_0 = 1$ be the initial approximation to the solution of $f(x) = 0$, using Newton's method. Find x_{100} (the approximation to the root of the equation $f(x) = 0$ after 100 iterations).

Solution: We won't solve this problem by actually calculating the iterations, but rather by some straight-forward reasoning. Since f is a linear function, its tangent line coincides with the graph of f. Hence, x_1 will already be the zero of f, namely, $x_1 = 5$.

3.6   Suppose that $f(r) = 0$, and that $f'(x)$ and $f''(x)$ are positive in the interval from $x = r$ to $x = x_0 > r$. Show that in this case $x_0 > x_1 > x_2 > \ldots > r$, so that the successive approximations converge *down* to r.

Solution: Since f and f' are positive in $[r, x_0]$, the curve $y = f(x)$ is *increasing* and *concave up* on that interval. Hence, the tangent to the curve at the point $(x_0, f(x_0))$ lies *below* the curve. (Draw a sketch.) As a result, the tangent line intersects the x-axis somewhere between r and x_0, so that $r < x_1 < x_0$. Now repeat the same argument with x_1 and x_2, obtaining $r < x_2 < x_1 < x_0$, and continue in the same fashion to complete the proof.

LINEAR APPROXIMATION AND TAYLOR POLYNOMIALS

3.7   **a.**   Compute the tangent line approximation to $f(x) = \sqrt{x}$ at $a = 100$ (Figure 3-6, page 63).

b.   Use this result to estimate $\sqrt{101}, \sqrt{102}$, and $\sqrt{105}$.

c.   What bound do you obtain on the errors, $E_1(101), E_1(102)$ and $E_1(105)$?

d.   Even though 101 is exactly as far away from 100 as 26 is from 25, note that the the error bound $E_1(101)$ is much smaller than $E_1(26)$. Moreover, the actual error, $P_1(101; 100) - \sqrt{101}$, is also smaller than the corresponding error, $P_1(26; 25) - \sqrt{26}$. (Recall the notation $P_1(x; a)$, introduced on page 62.) Similar results are true for 27 and 102, each 2 away from its base point, and for 30 and 105, each 5 away from its base point. Explain why the tangent line approximation is better near 100 than near 25 by examining the graph of $y = \sqrt{x}$ (Figure 3-6, page 63).

Solution:

a.   For $f(x) = \sqrt{x}$, $f'(x) = 1/(2\sqrt{x})$, so that $f(100) = \sqrt{100} = 10$ and $f'(100) = 1/20 = .05$. Hence

$$P_1(x) = f(100) + f'(100)(x - 100) = 10 + .05(x - 100).$$

b. $P_1(101) = 10 + .05(101 - 100) = 10.05$, $P_1(102) = 10 + .05(102 - 100) = 10.1$, and $P_1(105) = 10 + .05(105 - 100) = 10.25$.

c. $f''(x) = -1/(4x^{-3/2})$, so that

$$E_1(x) = \frac{f''(c)(x - 100)^2}{2} = -\frac{1}{8c^{-3/2}}(x - 100)^2,$$

where c lies between 100 and x. Hence,

$$|E_1(101)| = \left|-\frac{1}{4c^{3/2}}\right| \le \frac{1}{8 \cdot 10^{3/2}} = \frac{1}{8000} = .000125.$$

Similarly, $|E_1(102)| \leq 4/8000 = .0005$ and $E_1(105) = 25/8000 = .003125$.

d. The efficiency of the approximation is determined by how quickly the curve bends near the base point, a. Now, the bending of the curve is controlled by the second derivative, and we saw that $f''(x)$ is about $.000125$ near 100, while near 25 it is about 8 times as large.

3.8 **a.** Derive the tangent line approximation to $\sqrt[3]{x} = x^{1/3}$ at $a = 27$, and use it to estimate $\sqrt[3]{30}$.

b. Potentially how large is the error in this estimate?

c. Compare your estimate obtained here with the value of $\sqrt[3]{30}$ given by a calculator, and show that the error is smaller than the predicted bound.

Solution:

a. For $f(x) = x^{1/3}$, we have $f'(x) = x^{-2/3}/3$, so that $f(27) = 3$ and $f'(27) = 1/27$. Hence,

$$P_1(x) = 3 + \frac{1}{27}(x - 27),$$

so that $P_1(30) = 3 + 1/9 = 3.111111111$.

b. With $27 < c < 30$, we have

$$E_1(x) = \frac{f''(c)(x - 27)^2}{2} = -\frac{(x - 27)^2}{9c^{5/3}},$$

so that

$$|E_1(30)| = \frac{3^2}{9c^{5/3}} \leq \frac{1}{243} = .004115226.$$

c. From a calculator $\sqrt[3]{30} = 3.107232506$. The actual error is $.003878605$, which is somewhat smaller than the error bound of $.004115226$.

3.9 **a.** Compute $P_2(x; 27)$ and $P_3(x; 27)$ for $\sqrt[3]{x}$.

b. Use these results to estimate $\sqrt[3]{30}$.

c. Compare these results with the estimate of $\sqrt[3]{30}$ obtained using Newton's method in Solved Problem 3.1.

Solution:

a. We need to know $f''(27)$ and $f'''(27)$ in order to compute P_2 and P_3. Since

$$f'(x) = \frac{1}{3}x^{-2/3},$$

we have

$$f''(x) = -\frac{2}{9}x^{-5/3} \quad \text{and} \quad f'''(x) = \frac{10}{27}x^{-8/3},$$

so that $f''(27) = -2/2187$ and $f'''(27) = 10/177147$. Thus

$$P_2(x) = P_1(x) + \frac{f''(27)(x-27)^2}{2} = 3 + \frac{1}{27}(x-27) - \frac{(x-27)^2}{2187}.$$

Similarly,

$$P_3(x) = P_2(x) + \frac{f'''(x)(x-27)^3}{6} = 3 + \frac{1}{27}(x-27) - \frac{(x-27)^2}{2187} + \frac{5(x-27)^3}{531441}.$$

b.   Substituting $x = 30$, we obtain $P_2(30) = 3.106995885$ and $P_3(30)) = 3.107249911$.

c.   Using Newton's method we obtained 9 place accuracy with just 3 iterations. Here, the potential errors are much larger, with $E_2(30) \le .000236621$ and $E_3(30) \le .000017405$. As we've seen earlier, Newton's method, *when it works*, is exceptionally rapid.

3.10   In Example 3.7 we estimated $\sqrt{35}$ from $P_3(x; 25)$ and $P_4(x; 25)$. Since 35 is some distance from 25, a more efficient way of calculating $\sqrt{35}$ is to use a perfect square closer to 35 as the base point. Using $a = 36$, find the approximations to $\sqrt{35}$ obtained from P_n for $n = 1, 2, 3$, and 4.

Solution: We'll need the first 4 derivatives of $f(x) = \sqrt{x}$ at the base point 36. A straight-forward calculation yields $f'(36) = 1/12$, $f''(36) = -1/1728$, $f'''(36) = 1/124416$ and $f''''(36) = -5/35831808$. Also, $f(36) = \sqrt{36} = 6$. Thus,

$$P_1(x) = f(36) + f'(36)(x-36) = 6 + \frac{(x-36)}{12}$$

$$P_2(x) = P_1(x) + \frac{f''(36)(x-36)^2}{2} = 6 + \frac{(x-36)}{12} - \frac{(x-36)^2}{1728}$$

$$P_3(x) = P_2(x) + \frac{f'''(36)(x-36)^3}{6} = 6 + \frac{(x-36)}{12} - \frac{(x-36)^2}{1728} + \frac{(x-36)^3}{124416}$$

$$P_4(x) = P_3(x) + \frac{f''''(36)(x-36)^4}{24}$$

$$= 6 + \frac{(x-36)}{12} - \frac{(x-36)^2}{1728} + \frac{(x-36)^3}{124416} - \frac{5(x-36)^4}{35831808}$$

Substituting $x = 35$ into the above we obtain $P_1(35) = 5.916667$, $P_2(35) = 5.916088$, $P_3(35) = P_4(35) = 5.916080$. The last two are correct to 6 decimal places, so we've obtained a much better approximation than when we used $P_n(x; 25)$.

3.11   Find $P_n(x; 0)$ for $f(x) = e^x$.

Solution: This is one of the simplest Taylor polynomials to compute, since all the derivatives of f are equal to e^x. Hence $f(0) = f'(0) = f''(0) = \cdots = 1$, so that

$$P_n(x) = 1 + x + \frac{x^2}{2} + \cdots + \frac{x^n}{n!}.$$

3.12 Suppose we want to approximate $\sqrt{e}$ using Taylor polynomials, P_n, for e^x.

a. What is the smallest value of n which will guarantee that the error does not exceed .001?

b. Estimate $\sqrt{e}$ by evaluating $P_n(.5)$ for the n found in the first part of this problem.

Solution:

a. We saw in the previous problem that the Taylor polynomial of e^x of degree n about $a = 0$ is equal to

$$P_n(x) = 1 + x + \frac{x^2}{2!} + \cdots + \frac{x^n}{n!}.$$

The error is given by

$$E_n(x) = \frac{e^c \, x^{n+1}}{(n+1)!},$$

where c lies between 0 and x. We have to determine a value of n which will guarantee that $|E_n(.5)| < .001$. Since e^c is included in the error estimate, we need a bound on this term. While accurate estimates for e require more sophisticated analysis, we can use a cruder bound, namely, $e < 3$. Hence $e^c < e^{.5} < \sqrt{3} < 1.75$. Let's look at the error for, say, $n = 3$.

$$|E_3(.5)| < \frac{1.75\,(.5)^4}{4!} = .0046.$$

Since this error is unacceptably large, we increase n to 4, obtaining

$$|E_4(.5)| < \frac{1.75\,(.5)^5}{5!} = .00046 < .001.$$

Thus $n = 4$ suffices.

b.

$$P_4(.5) = 1 + \frac{1}{2} + \frac{1}{8} + \frac{1}{48} + \frac{1}{384} = 1.64844.$$

Since the exact value of $\sqrt{e} = 1.64872$ (to 5 decimal places), the actual error is .00028, which is well below the requirement.

3.13 Suppose that $f''(x) > 0$ for all x in some interval $[a - h, a + h]$ surrounding the point a. Show that the tangent line approximation at $(a, f(a))$ lies *below* the curve $y = f(x)$ throughout $[a - h, a + h]$. Hence, we see that in this case the tangent line *underapproximates* the function.

Solution:

$$f(x) = f(a) + f'(a)(x - a) + \frac{f''(c)(x - a)^2}{2} = P_1(x) + \frac{f''(c)(x - a)^2}{2}$$

for some c between a and x. If $x \in [a - h, a + h]$, then c also lies in this interval, so that $f''(c) > 0$. Since $(x - a)^2 \geq 0$ for all x, we see that $f(x) \geq P_1(x)$ for all $x \in [a - h, a + h]$.

Supplementary Problems

NEWTON'S METHOD

3.14 Solve the equation $e^x = 3x$ by Newton's Method. How many solutions are there?

3.15 Use Newton's method to approximate the following quantities:

  a. $\sqrt{43}$ **b.** $\sqrt[3]{15}$.

3.16 Find the root of the equation $\sin x = x - 1$, using Newton's method.

LINEAR APPROXIMATION AND TAYLOR POLYNOMIALS

3.17 Find P_1, P_2, and P_3 for the following functions at the indicated points:

  a. $\sin x$ at $a = 0$.
  b. $\cos x$ at $a = \pi/2$.
  c. $\sqrt{1 + x}$ at $a = 0$.
  d. $\ln(1 + x)$ at $a = 0$.

3.18 Find the error estimates, $E_3(x)$, for each of the functions in the previous problem.

3.19 **a.** Find the second order Taylor polynomial P_2 for the function $f(x) = \ln x$ at $a = 2$.

b. Find an expression for the error, $E_2(x)$.

c. Find an upper bound for the error when $P_2(x)$ is used to approximate $f(3) = \ln 3$.

Answers to Supplementary Problems

3.14 .6190612867 and 1.512134552.

3.15 **a.** $x_0 = 6$, $x_1 = 6.583333333$, $x_2 = 6.557489451$, $x_3 = 6.557438524$.

b. $x_0 = 2$, $x_1 = 2.583333333$, $x_2 = 2.471441785$, $x_3 = 2.466223133$,
$x_4 = 2.466212074$.

3.16 $x_0 = 1$, $x_1 = 2.830487722$, $x_2 = 2.049555245$, $x_3 = 1.938656127$,
$x_4 = 1.934568962$, $x_5 = 1.934563211$.

3.17 **a.** $P_1(x) = P_2(x) = x$, $P_3(x) = x - \dfrac{x^3}{6}$.

b. $P_1(x) = P_2(x) = -\left(x - \dfrac{\pi}{2}\right)$

$P_3(x) = -\left(x - \dfrac{\pi}{2}\right) + \dfrac{1}{6}\left(x - \dfrac{\pi}{2}\right)^3$

c. $P_1(x) = 1 + \dfrac{x}{2}$

$P_2(x) = 1 + \dfrac{x}{2} - \dfrac{x^2}{8}$

$P_3(x) = 1 + \dfrac{x}{2} - \dfrac{x^2}{8} + \dfrac{x^3}{16}$

d. $P_1(x) = x$

$P_2(x) = x - \dfrac{x^2}{2}$

$P_3(x) = x - \dfrac{x^2}{2} + \dfrac{x^3}{3}$

3.18 **a.** $E_3(x) = \dfrac{(\sin c)\, x^4}{24}$, for some c between 0 and x.

b. $E_3(x) = \dfrac{1}{24}(\cos c)\left(x - \dfrac{\pi}{2}\right)^4$, for some c between $\pi/2$ and x.

c. $E_3(x) = -\dfrac{5\,x^4}{128\,(1+c)^{7/2}}$, for some c between 0 and x.

d. $E_3(x) = -\dfrac{x^4}{4\,(1+c)^4}$, for some c between 0 and x.

3.19 a. $P_2(x) = \ln 2 + \dfrac{(x-2)}{2} - \dfrac{(x-2)^2}{8}$

b. $E_2(x) = \dfrac{(x-2)^3}{3\,c^3}$, for some c between 2 and x.

c. $|E_2(x)| \le \dfrac{1}{24}$.

Chapter 4

The Integral

4.1 Motivation

The second major concept of calculus is the integral, a topic perhaps even richer in applications than the derivative. The derivative concentrates on local features of functions, for example, the slope of a curve *at a point* (Figure 4-1), or the *instantaneous*

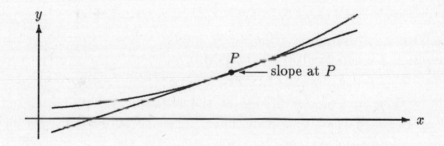

Figure 4-1: Slope of the curve at P

velocity of a moving body. The integral, on the other hand, is concerned with global properties, such as the *total* distance covered by a moving body, the area under a curve over some *interval* (Figure 4.2), and so forth. As we'll see, the integral displays once again one of the basic themes of calculus: The progression from simple, well-understood ideas to more sophisticated ones. Just as in Chapters 2 and 3, Approximation — Refinement — Limit ($\mathcal{A}$—$\mathcal{R}$—$\mathcal{L}$) will be the method we employ over and over as we develop important tools which are used to solve a variety of problems in many different areas. The examples that we'll explore include velocity and distance, force and work, and area, but there are many additional applications of the integral; we'll study some of them in Chapter 5.

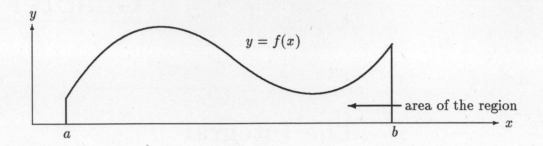

Figure 4-2: Area under $y = f(x)$ from a to b

4.1.1   Velocity and Distance

What We Know: The total distance traveled by a body moving with *constant* velocity.

What We Want To Know: The total distance traveled by a body moving with *non-constant* velocity.

How We Do It: Approximate the *non-constant* velocity function with velocity functions which are *constant on subintervals*.

As we have seen in Chapter 2, one of the fundamental problems of calculus is the connection between velocity and distance. The computation of the instantaneous velocity when we know the distance function was one of the key ideas which motivated the derivative. In this chapter, we turn things around: We use our knowledge of the velocity function in order to compute the distance. We will see, for example, that if the *odometer* of an automobile is broken, we can still compute the total distance traveled, provided the *speedometer* is working and that we have a complete record of our instantaneous velocity at each moment of our trip. We begin modestly, with the familiar formula, $d = rt$, valid for a body moving *along a straight line* with constant velocity. Our first step is to give a geometric representation of this formula.

Suppose a body moves along a straight line starting at time t_0 and ending its journey at time t_1, moving throughout with constant velocity, r. Then, according to the $d = rt$ formula, the distance covered is $r(t_1 - t_0)$. Consider, now, a rectangle of height r, extending from t_0 to t_1, whose area, of course, is equal to $r(t_1 - t_0)$ (Figure 4-3). Hence, the distance traveled is equal to the area of an appropriate rectangle.

Now suppose that the velocity of the body changes once, so that the velocity is r_1 from time t_0 to time t_1 and r_2 from t_1 to t_2 (Figure 4-4). It is clear that the total

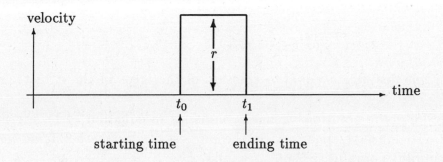

Figure 4-3: Motion with constant velocity

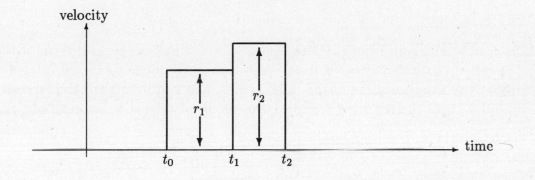

Figure 4-4: Velocity changes once

distance covered is $r_1(t_1 - t_0) + r_2(t_2 - t_1)$. We also see from Figure 4-4 that this distance is equal to the sum of the areas of the two rectangles.

It is easy to extend this result to a situation in which the velocity of the body changes several times. Suppose that the velocity is r_1 from time t_0 to time t_1, the velocity is r_2 from t_1 to t_2, and so forth, with the final velocity being r_n from t_{n-1} to t_n (Figure 4-5). (The function whose graph is displayed in Figure 4-5 is called a

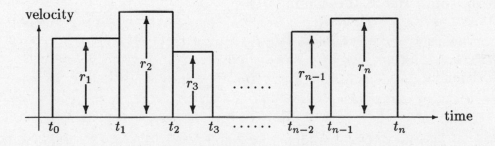

Figure 4-5: Several changes in velocity

step-function.) Then the total distance covered is equal to

$$r_1(t_1 - t_0) + r_2(t_2 - t_1) + \cdots + r_n(t_n - t_{n-1}). \tag{4.1}$$

Once again, this distance is equal to the sum of the areas of the n rectangles seen in Figure 4-5.

An aside for some notation: Lengthy sums, as in equation (4.1), are a feature of the integral. To conserve space and in the interest of clarity, the *sigma notation* was devised to simplify complicated sums, as follows.

Instead of writing a sum such as $a_1 + a_2 + \cdots + a_n$, we write

$$\sum_{i=1}^{n} a_i.$$

This is read: "The sum from $i = 1$ to n of a sub-i." a_i is the general term of the sum and i is called the *index of summation*. We successively replace i by *all* integer values starting at the smallest value of the index, $i = 1$, and continuing till the largest value, $i = n$, and add up the resulting terms. This shorthand tells you, for example, that the sum

$$1 + \frac{1}{2} + \frac{1}{3} + \cdots + \frac{1}{100}$$

can be written as

$$\sum_{i=1}^{100} \frac{1}{i},$$

while the sum

$$1 + 3 + 5 + \cdots + 29$$

is abbreviated to

$$\sum_{i=1}^{15} (2i - 1).$$

(**Note:** Every odd number can be written in the form $2i - 1$, and there are exactly 15 terms in the sum.) Hence, the sum in (4.1),

$$r_1(t_1 - t_0) + r_2(t_2 - t_1) + \cdots + r_n(t_n - t_{n-1})$$

can be written as

$$\sum_{i=1}^{n} r_i(t_i - t_{i-1}). \tag{4.2}$$

This notation can be further simplified by defining $\Delta t_i = t_i - t_{i-1}$, thereby reducing (4.2) to

$$\sum_{i=1}^{n} r_i \Delta t_i. \tag{4.3}$$

Remark 4.1   The letter i used for the index is called a 'dummy variable.' i can be replaced by any other letter, such as k. Thus $\sum_{i=1}^{n} a_i$ and $\sum_{k=1}^{n} a_k$ are *both* shorthand for the sum $a_1 + a_2 + \cdots + a_n$. We could also write $\sum_{j=1}^{n} a_j$ or $\sum_{l=1}^{n} a_l$ to represent this sum.

Exercise 4.1   *Evaluate:*

$$\text{a. } \sum_{i=1}^{8} i^2; \quad \text{b. } \sum_{i=1}^{5} \frac{1}{i(i+1)}.$$

Exercise 4.2   *Abbreviate the following sums using the sigma notation:*
a. $2 + 4 + 6 + \cdots + 18 + 20;$ **b.** $a_1 b_1 + a_2 b_2 + \cdots + a_n b_n.$

For more details on the sigma notation, consult your textbook.

We have now arrived at the main problem, the transition from step functions to arbitrary continuous functions. Suppose the velocity function of a body moving along a straight line is $v = r(t)$, $\quad a \leq t \leq b$, (Figure 4-6). What is the total distance

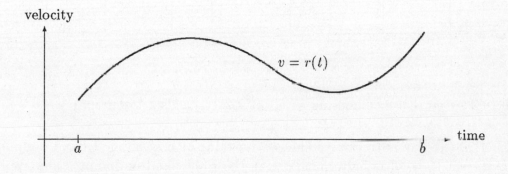

Figure 4-6: Graph of a velocity function

covered by the body? We are not yet in a position to give an exact answer to this question. Instead, in line with our general approach to problems in calculus, we begin with an *approximation*. We replace the function $v = r(t)$ with an appropriately chosen step-function, obtained as follows: Select a number of points in the interval $[a, b]$, and label them $t_0, t_1, \ldots, t_n$, where $t_0 = a$ and $t_n = b$. We now pretend that the velocity of the body is constant in each of the intervals, $[t_{i-1}, t_i], i = 1, 2, \ldots, n$. But which constant should it be? There are many ways of choosing the values of the approximate velocity on the intervals, $[t_{i-1}, t_i]$. We could, for example, choose the largest value of the function, r, in each of the subintervals. This choice gives a so-called *upper sum*, which is greater than the actual distance traveled. Or, we could choose the smallest such value, to obtain a *lower sum*, which is smaller than the actual value. These choices lead to the pictures seen in Figures 4-7 and 4-8, respectively.

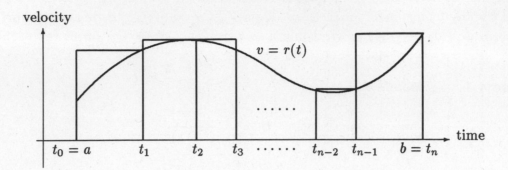

Figure 4-7: Choosing largest values yields an upper sum

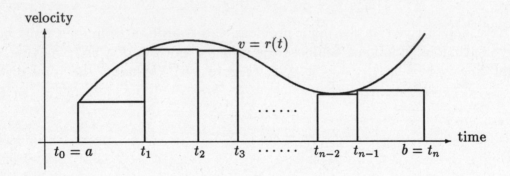

Figure 4-8: Choosing smallest values yields a lower sum

Another possibility is to always choose the value of r at the left-hand endpoint of each subinterval $[t_{i-1}, t_i]$ or, alternatively, at the right-hand endpoint. The approach that we adopt, however, is to *arbitrarily* choose a point c_i *somewhere* in the interval $[t_{i-1}, t_i]$, and replace r by $r(c_i)$ in $[t_{i-1}, t_i], i = 1, 2, \ldots, n$ (Figure 4-9).

Remark 4.2 By "arbitrarily," we mean that the method of choice of c_i is not specified. We *could* pick the left-hand endpoint of each subinterval; *or*, as in Figure 4-7, the point at which r achieves its maximum value in the subinterval; *or* we could choose c_i through some random process, say by 'throwing a dart' at the subinterval. It will turn out that although the method used to choose c_i will generally influence the intermediate computations, it will have no effect on the ultimate outcome. The proof of this result is beyond the scope of this work, but can be found in textbooks on advanced calculus.

The step-function we have obtained is an *approximation* to the velocity function, r. Hence, by Equation (4.3), the total distance covered is approximately equal to

$$\sum_{i=1}^{n} r(c_i) \Delta t_i. \tag{4.4}$$

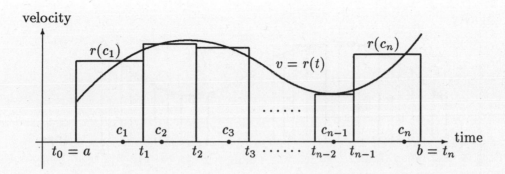

Figure 4-9: Choosing arbitrary points, c_i

As before, expression (4.4) has a geometric interpretation: It is the sum of the areas of the rectangles in Figure 4-9.

We now *refine* our approximation by further subdividing the interval $[a, b]$ into a larger number of subintervals. We expect that as the number of subintervals grows, the *approximate* distance covered comes ever closer to the *exact* distance covered by the moving body.

Remark 4.3 As opposed to the situation with the derivative, where for some functions (such as $|x|$) the refinement process can fail to improve the approximation, the integral is much better behaved. Only for severely 'pathological' functions (which will never occur on an exam!) does the procedure fail. In particular, it works for all *continuous* functions.

As we continue to refine the approximation, Figure 4-9 is superseded by Figures 4-10 and 4-11.   Now, to obtain the exact value of the distance covered, we pass to the *limit*,

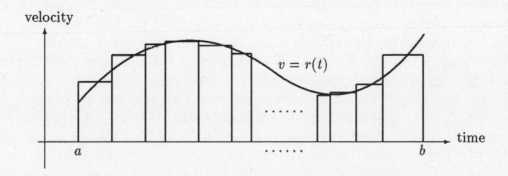

Figure 4-10: Refined approximation

In other words, we imagine that the process of subdividing the interval $[a, b]$ continues

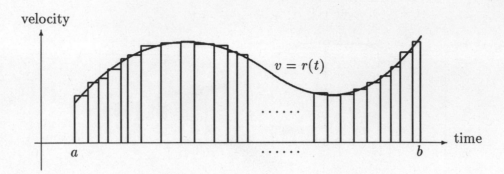

Figure 4-11: Further refinement

indefinitely. Thus, the total distance covered is equal to

$$\lim \sum_{i=1}^{n} r(c_i) \Delta t_i. \qquad (4.5)$$

(The precise definition of the limit in (4.5) will come in Section 4.2.) But what is the *geometric* interpretation of the exact distance covered? There's nothing more natural than to say that the total distance is *equal to the area under the curve* $v = r(t)$, for $a \le t \le b$.

Up to this point, the only case we have considered is one in which the velocity function, r, is *positive* on the interval, $[a, b]$. In this case, the approximating rectangles all lie above the t-axis and, as we saw, the sum of their areas is an approximation to the total distance covered by the moving body (Figure 4-9, page 95). We now wish to allow r to be *negative*, as well.

But what can we possibly mean by a negative velocity function? (No, it does not mean that the body is moving backwards in time!) To understand this idea of negative velocity, let's distinguish *velocity* from *speed*. The *speed* of a body tells only how fast it is moving, while the *velocity* also specifies its *direction*. (In technical terminology, speed is a *scalar* quantity, having only magnitude, while velocity is a *vector* quantity, having both magnitude and direction.) The distinction between speed and velocity is important: If you are caught driving 80 miles per hour, then your fine for speeding will be the same regardless of the direction you were heading. On the other hand, if you are going 50 miles per hour and collide with another car going 45 miles per hour, then it *certainly does matter* whether the two cars were heading in the *same* or the *opposite* direction.

Now, how is the direction of motion specified? (Recall that, in this section, we are discussing only motion along a *straight line.*) We choose one direction along the line, say northward, to be the positive one: A body moving in that direction will be said to have *positive* velocity. If the body moves in a southward direction, its velocity will be *negative* (Figure 4-12).

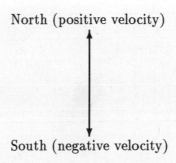

Figure 4-12: Direction of motion

Now, how does all of this fit in to what we've done thus far? Let's look at the situation geometrically, again. Suppose a car leaves home and travels North for 3 hours with a velocity of 50 miles per hour. It then turns around and travels South for 2 hours, with a velocity of -40 miles per hour (Figure 4-13). Where is the car at the end of this

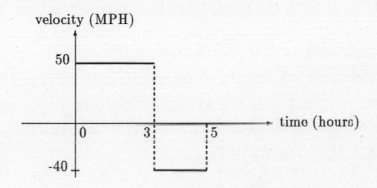

Figure 4-13: Displacement

5-hour trip? Well, it moved $3 \cdot 50 = 150$ miles in the positive direction (North), and then came back *towards its home* (moving South) a total of $2 \cdot 40 = 80$ miles. So the car's *displacement*—how far it is from home—is $150 - 80 = 70$ miles. We see, that in computing how far we are from home (the starting point), we must *subtract*. In other words, the displacement is equal to

$$3 \cdot 50 + 2 \cdot (-40) = 150 - 80 = 70 \text{ miles.}$$

Notice, that we are talking about the displacement from the starting point, that is, the distance from home, and *not* the *total distance* covered by the car, which is equal to

$$3 \cdot 50 + 2 \cdot |-40| = 150 + 80 = 230 \text{ miles.}$$

So in computing the displacement, we consider the area of the second rectangle in Figure 4-13 to be *negative*, while that of the first is *positive*. In the general case, we *add* the areas of rectangles lying above the t-axis (those with positive velocity). From this sum, we then *subtract* the sum of the areas of those rectangles lying below the t-axis (those with negative velocity). Hence, the sum,

$$\sum_{i=1}^{n} r(c_i)\Delta t_i$$

is an *approximation* to the *displacement* of the body from its starting point, and the exact value of the displacement is equal to

$$\lim \sum_{i=1}^{n} r(c_i)\Delta t_i. \tag{4.6}$$

On the other hand, the *total distance* traveled by the body is given by

$$\lim \sum_{i=1}^{n} |r(c_i)|\Delta t_i. \tag{4.7}$$

(4.7) is true because $|r(t)|$ is a positive function, bringing us back to our original situation (4.5).

4.1.2   Work

We turn now to another physical problem which will help motivate the integral.

What We Know: The work done by a *constant* force acting on a body.

What We Want To Know: The work done by a *non-constant* force acting on a body.

How We Do It: Approximate the *non-constant* force by forces which are *constant on subintervals.*

Suppose a *constant* force, F, is applied in moving a body along a straight line from a point x_0 to another point x_1, say pushing a desk across a room. Then the *work* done, W, is defined to be the *product* of the force and the distance covered. Symbolically,

$$W = F \times (x_1 - x_0).$$

For example, if you push the desk with a constant, *unchanging* force $F = 10$ over a distance of 8, then the work will be $10 \times 8 = 80$. Let us again translate this formula

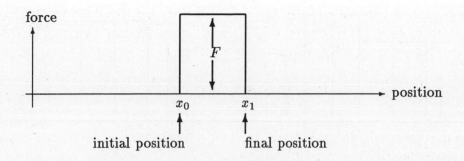

Figure 4-14: Work when force is constant

geometrically (Figure 4-14). As before, the work is given by the area of a rectangle, whose height this time is F and whose base is $x_1 - x_0$.

Now turn back to Figure 4-3 (page 91), where we were dealing with distance rather than work, and notice that it is identical to Figure 4-14, except for a change of letters and the labeling of the axes. Let's go on. It usually isn't possible for a person to push a desk with a *constant* force for a long distance. A more realistic situation is one in which for a while the applied force is one constant, F_1, and another constant, F_2, thereafter. The total work done is now $F_1(x_1 - x_0) + F_2(x_2 - x_1)$. This leads to Figure 4-15, which looks suspiciously like Figure 4-4 (page 91).

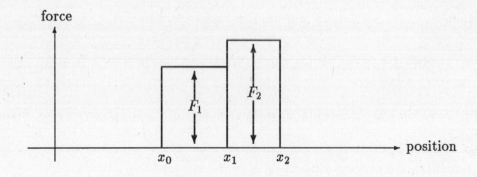

Figure 4-15: Force changes once

Now suppose that the desk has to be pushed for a very long distance, which is tiring. The force behind the pushes might now look like Figure 4-16, with the total work equal to

$$\sum_{i=1}^{n} F_i \Delta x_i. \tag{4.8}$$

Just as in our earlier discussion of distance, the work done has the geometrical significance of being equal to the sum of the areas of the rectangles in Figure 4-16.

The extension to the general case is now clear (Figure 4-17). When the force function

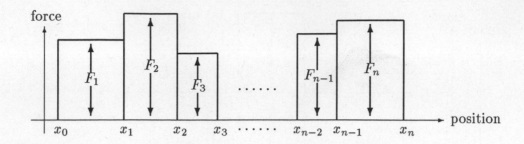

Figure 4-16: Several changes of force

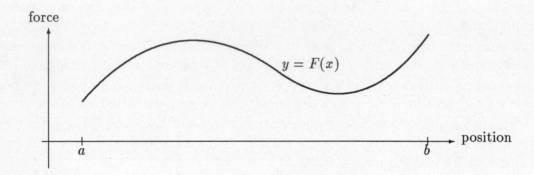

Figure 4-17: Force function

is a curve, we proceed as in Section 4.1.1. We *approximate* F by a step-function obtained in the following way:

1. We partition the interval $[a, b]$ with a set of points, $x_0, x_1, x_2, \ldots, x_n$ (where $x_0 = a$ and $x_n = b$);

2. We arbitrarily choose a point c_i in $[x_{i-1}, x_i]$, and

3. We let the step-function take on the value $F(c_i)$ on this subinterval (Figure 4-18).

Hence, the work is approximately equal to

$$\sum_{i=1}^{n} F(c_i)\Delta x_i. \tag{4.9}$$

We next *refine* the approximation by further subdividing the interval $[a, b]$ (see Figure 4-19). To obtain better and better approximations to the actual work done we continue the process of subdividing the interval $[a, b]$ (Figure 4-20).

But how do we obtain the exact value of the work done? For this, we must pass to the *limit*, which is done by continuing the process of subdivision indefinitely. By doing

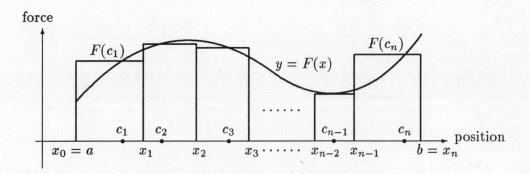

Figure 4-18: Approximating the force function

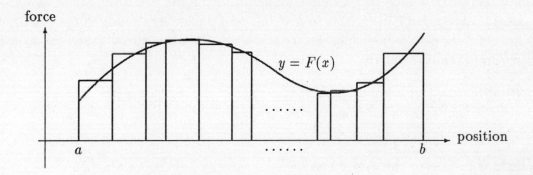

Figure 4-19: Refined approximation

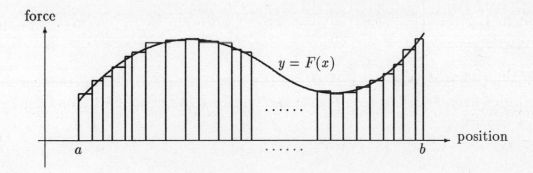

Figure 4-20: Further refinement

so, we find that the total work done by the force function $y = F(x)$ in pushing the desk is equal to

$$\lim \sum_{i=1}^{n} F(c_i)\Delta x_i. \tag{4.10}$$

$\mathcal{L}$

Once again, it is natural for us to associate this limit with the *area under the curve*, $y = F(x)$, $a \leq x \leq b$.

As in the previous section, we now consider force functions which are either negative or of variable sign. (Force, like velocity, is a *vector* quantity. The force of lifting a book is a *positive* one, but the force of gravity, which pulls the book down is a *negative* one.) In this case, however, there is no change in the definition: We define the work done on the body by (4.10).

4.1.3   Area

We have seen that two different problems led to the same solution. Both total distance and total work have been shown to be equal to the area under a curve. While this may seem like a nice way to solve these problems, we are a bit taken aback when we realize that we *don't know how to find the area under very many curves!* Unless the function is linear or a semicircle, we are stymied. We have, in fact, arrived at a third and very important problem:

Find the area under $y = f(x)$, $a \leq x \leq b$.

What We Know: Area of a *rectangle*.

What We Want To Know: Area of a region of the plane under an *arbitrary curve*.

How We Do It: Approximate the *arbitrary* region with a set of *rectangles*.

There is no need to go through the various stages of approximation, refinement and limit, since it is clear how to proceed from our work in Sections 4.1.1 and 4.1.2. The area is equal to

$$\lim \sum_{i=1}^{n} f(c_i) \Delta x_i. \tag{4.11}$$

Equation (4.11) assumes, of course, that the function f is *positive* on the interval $[a, b]$. If f can also assume negative values, then we no longer speak about the area *under* the curve, but rather about the area *between* the curve and the x-axis. This area is defined to be

$$\lim \sum_{i=1}^{n} |f(c_i)| \Delta x_i. \tag{4.12}$$

(4.12) is analogous to (4.7) for calculating the total distance.

Let's take stock of the situation. The expression

$$\lim \sum_{i=1}^{n} f(c_i)\Delta x_i$$

has now appeared in three separate contexts, first in connection with velocity and distance (4.5), then in our study of work (4.10), and finally, here, where we considered area (4.11). As an idea that keeps showing up, isn't it time to give it a name? We turn to that task in the next section.

4.2   Definition of the Integral

Let f be a bounded function on the interval $[a, b]$. (**Note:** f is bounded on $[a, b]$ if there exists a number, M, such that $|f(x)| \leq M$ for all x in $[a, b]$.) Let $a = x_0 < x_1 < x_2 < \ldots < x_n = b$ be a set of points which partitions the interval $[a, b]$, and let c_i be an arbitrary point in the subinterval $[x_{i-1}, x_i]$, $i = 1, 2, \ldots, n$. The expression

$$\lim \sum_{i=1}^{n} f(c_i)\Delta x_i \tag{4.13}$$

is called the *definite integral* of $f(x)$ on $[a, b]$.

We've been somewhat vague up to this point about the limit in (4.13), the nature of which is different from the one that we encountered in connection with the derivative. From our earlier discussion in Section 4.1 we have a fairly good intuitive idea of the process: The subdivision of the interval $[a, b]$ continues indefinitely, by adding more and more points. So it seems plausible to say that the limit in (4.13) consists of letting n (the number of subintervals) tend to infinity. Well, this idea may be plausible, but, unfortunately, it is also wrong! For consider the picture in Figure 4-21. The number

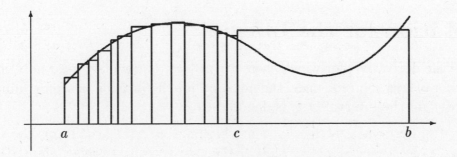

Figure 4-21: Partially refined partition

of partition points (or the number of rectangles) is very large here. Between a and c, the approximation of the function by the step-function (and of the area by the sum of the areas of the rectangles) is excellent. But over the entire interval, $[a, b]$, the approximation is not very good, because it is not sufficiently *refined*. In other words, the refinement stage here consists not merely of increasing the *number* of rectangles. We must also make sure that the partition is not a coarse one. We accomplish this as follows: Denote by P the partition of $[a, b]$. That is, P is the set consisting of the points $\{x_0, x_1, \ldots, x_n\}$. We now define the *mesh* of the partition, denoted by $||P||$, to be the *largest* of the subintervals. Thus, $||P|| = \max(x_i - x_{i-1})$, $i = 1, 2, \ldots, n$. The key to the refinement process, which causes the approximation to improve (at least for reasonable functions), is to make sure that $||P||$ tends to 0. Note that in Figure 4-21, $||P||$ is equal to $(b - c)$, which is large, even though the other subintervals are small, so that the partition is coarse. The complete definition of the definite integral becomes the following:

$$\lim_{||P|| \to 0} \sum_{i=1}^{n} f(c_i) \Delta x_i, \tag{4.14}$$

where x_i and c_i have the same meaning as before.

Note that a simple way to guarantee that $||P||$ goes to 0 is to insist that the subintervals in the partition be equally spaced, so that $x_i - x_{i-1} = (b - a)/n$. In this case, $||P|| \to 0$ *is* equivalent to $n \to \infty$. However, there are situations in which the flexibility of a nonuniform partition is useful, so we allow for it in the definition.

Remark 4.4 The definite integral defined above is also known as the Riemann integral (named after the 19th century mathematician, George Bernhard Riemann), and the sums (4.4), $\sum_{i=1}^{n} r(c_i)(t_i - t_{i-1})$, or (4.9), $\sum_{i=1}^{n} F(c_i)(x_i - x_{i-1})$, are called Riemann sums. Hence, we see from (4.14) that the Riemann integral of a function is the limit, as the mesh of the partition tends to 0, of Riemann sums of the function. Many of the applications of the integral depend upon our being able to recognize specific Riemann sums.

Before considering some examples, we turn to the notation for the integral.

4.3   Notation for the Integral

As opposed to the derivative, for which there are at least two widely used notations, as well as a number of less common ones, the notation for the integral is almost universal. The integral of f on the interval $[a, b]$ is denoted by

$$\int_a^b f(x) \, dx.$$

Since the integral is the limit of the Riemann sums (4.14), we have, symbolically,

$$\int_a^b f(x) \, dx = \lim_{||P|| \to 0} \sum_{i=1}^{n} f(c_i) \Delta x_i.$$

4.3.1   Riemann Sums

Let's use this notation in conjunction with Riemann sums. In all of the following formulas it is assumed that f is defined on an interval $[a, b]$, that there are partitions of the interval, P, given by $a = x_0 < x_1 < \ldots < x_n = b$, and that $c_i \in [x_{i-1}, x_i]$, $i = 1, 2, \ldots, n$.

For example, a Riemann sum for the function $f(x) = x^2$ on the interval $[a, b]$ is given by $\sum_{i=1}^{n} (c_i)^2 \Delta x_i$. Hence, the integral of x^2 on this interval,

$$\int_a^b x^2 \, dx = \lim_{\|P\| \to 0} \sum_{i=1}^{n} (c_i)^2 \Delta x_i.$$

Similarly,

$$\lim_{\|P\| \to 0} \sum_{i=1}^{n} \frac{c_i}{1 + c_i} \Delta x_i = \int_a^b \frac{x}{1 + x} \, dx,$$

since for $f(x) = x/(1 + x)$ we have $f(c_i) = c_i/(1 + c_i)$. Other examples of Riemann sums and their limits are:

$$\lim_{\|P\| \to 0} \sum_{i=1}^{n} \sin c_i \Delta x_i = \int_a^b \sin x \, dx$$

$$\lim_{\|P\| \to 0} \sum_{i=1}^{n} \pi (f(c_i))^2 \Delta x_i = \int_a^b \pi (f(x))^2 \, dx$$

$$\lim_{\|P\| \to 0} \sum_{i=1}^{n} 2\pi c_i f(c_i) \Delta x_i = \int_a^b 2\pi x f(x) \, dx$$

$$\lim_{\|P\| \to 0} \sum_{i=1}^{n} \sqrt{1 + (f'(c_i))^2} \, \Delta x_i = \int_a^b \sqrt{1 + (f'(x))^2} \, dx.$$

The last three examples were not chosen randomly; they all involve important applications of the integral, and our ability to recognize these and other Riemann sums is the key to the solution of these problems.

> *Often, when a sum includes a term Δx_i, it either is, or can be manipulated into, a Riemann sum.*

Exercise 4.3   *Write down Riemann sums for the following functions:*
a. $f(x) = x^2 + 3x + 1$ *on the interval* $[a, b]$.
b. $f(x) = x + \cos x$ *on the interval* $[0, \pi]$. *What are* x_0 *and* x_n *in this case?*

Exercise 4.4 *Express the following limits of Riemann sums as integrals (here P is a partition of $[0,1]$).*

$$\textbf{a.} \quad \lim_{\|P\|\to 0} \sum_{i=1}^{n} (c_i^4 + 3c_i^2 - 2)\Delta x_i; \quad \textbf{b.} \quad \lim_{\|P\|\to 0} \sum_{i=1}^{n} \frac{c_i^2 + 1}{c_i^2 + 2c_i + 4}\Delta x_i.$$

Remark 4.5 A shortcoming of the notation for a Riemann sum is the lack of indication of the underlying interval, $[a, b]$. It is, however, understood that such an interval exists, and that P is a partition of that interval with $x_0 = a$ and $x_n = b$.

Let's try to get a feel for how we apply the definition of the definite integral. We'll begin with a very simple example, one for which we already know the answer without the use of calculus, but one which will be quite instructive, nevertheless.

Example 4.1 *Find the area under the line $y = x$, $0 \le x \le 1$ (Figure 4-22).*

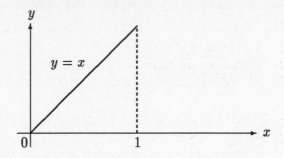

Figure 4-22: Graph of $y = x$

Solution: The area under this straight line is clearly 1/2, but let's go through the calculation using the definition (4.14),

$$\lim_{\|P\|\to 0} \sum_{i=1}^{n} f(c_i)\Delta x_i.$$

We choose the partition P of $[0,1]$ to be an equally spaced one. Hence, with n points, $x_0 = 0$, $x_1 = 1/n$, $x_2 = 2/n$, and, in general, $x_i = i/n$. Thus $\Delta x_i = 1/n$, and

$$\lim_{\|P\|\to 0}$$

is equivalent to

$$\lim_{n\to\infty}.$$

For the points c_i we choose the right-hand endpoint of each subinterval, $c_i = i/n$. (Recall that the choice of c_i is at our discretion.) Thus, (Figure 4-23)

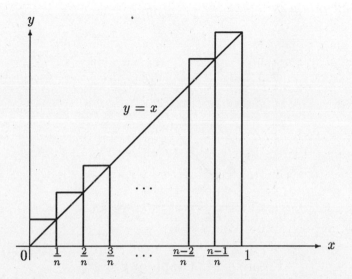

Figure 4-23: Choosing right-hand endpoints

$$\int_0^1 f(x)\,dx \quad = \quad \int_0^1 x\,dx$$

$$= \quad \lim_{\|P\|\to 0} \sum_{i=1}^{n} c_i \Delta x_i$$

$$= \quad \lim_{n\to\infty} \sum_{i=1}^{n} \frac{i}{n}\frac{1}{n}$$

$$= \quad \lim_{n\to\infty} \frac{1}{n^2} \sum_{i=1}^{n} i \qquad (4.15)$$

In order to complete the solution of the problem, we need to know how to evaluate the last sum. (Notice how complicated things are getting, and all we're trying to do is to find the area of a triangle!) Fortunately, sums of this type have been computed, using techniques that have nothing to do with calculus. It is known that

$$\sum_{i=1}^{n} i = \frac{n(n+1)}{2}. \qquad (4.16)$$

(This is simply the sum of an arithmetic progression.) Substituting (4.16) into (4.15) yields

$$\int_0^1 x\,dx = \lim_{n\to\infty} \frac{n(n+1)}{2n^2} = \frac{1}{2}.$$

Well, that's a great deal of work to verify a result that we already know! Imagine what will happen when we tackle something *new*! Let's try.

Example 4.2 *Find the area under the parabola $y = x^2$, for x between 0 and 1 (Figure 4-24).*

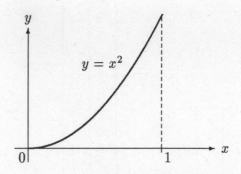

Figure 4-24: Graph of $y = x^2$

We proceed in a manner similar to Example 4.1, choosing an equally spaced partition consisting of the n points $x_i = i/n$, and again letting $c_i = i/n$ (Figure 4-25). We obtain

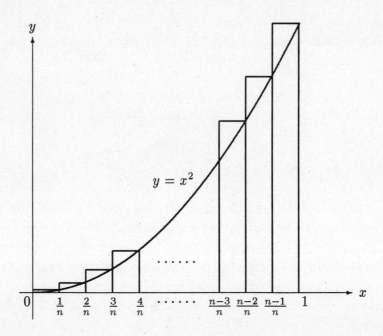

Figure 4-25: Choosing right-hand endpoints

$$
\int_0^1 x^2 \, dx = \lim_{n \to \infty} \sum_{i=1}^{n} \left(\frac{i}{n} \right)^2 \frac{1}{n}
$$

$$
= \lim_{n \to \infty} \frac{1}{n^3} \sum_{i=1}^{n} i^2.
$$

To evaluate this expression we use a formula similar to (4.16), namely, $\sum_{i=1}^n i^2 = n(n+1)(2n+1)/6$ which is equal to $(2n^3 + 3n^2 + n)/6$. Substitution yields

$$
\begin{aligned}
\int_0^1 x^2 \, dx &= \lim_{n \to \infty} \frac{1}{n^3} \sum_{i=1}^n i^2 \\
&= \lim_{n \to \infty} \frac{2n^3 + 3n^2 + n}{6n^3} \\
&= \lim_{n \to \infty} \frac{2n^3 \left(1 + \frac{3}{2n} + \frac{1}{2n^2}\right)}{6n^3} \\
&= \lim_{n \to \infty} \left(1 + \frac{3}{2n} + \frac{1}{2n^2}\right) / 3 \\
&= \frac{1}{3}
\end{aligned}
$$

since

$$
\lim_{n \to \infty} \frac{1}{n} = \lim_{n \to \infty} \frac{1}{n^2} = 0.
$$

So we have succeeded in getting a *new* result, but look at the price we've paid in terms of the work involved! It seems unlikely that this approach will be of much practical use, *unless* we can develop a method which allows us to circumvent this clumsy definition. This development will take place in the next section.

4.4   Computational Techniques

It is clear from the examples we examined in the previous section that the computation of integrals using the formal definition is exceptionally clumsy. But what can we do to remedy this situation? We have to search for alternative methods which are considerably simpler. Fortunately, this was done for us by Isaac Newton and his contemporaries more than 300 years ago. Known as the *Fundamental Theorem of Calculus*, its theoretical importance cannot be overstated, since it connects the two main concepts of calculus, the derivative and integral. For our purposes here, however, it can be thought of primarily as a shortcut, which allows us to compute integrals without using definition (4.14). We do not prove the fundamental theorem; instead, consult your text for a complete discussion and proof of this major result.

The Fundamental Theorem of Calculus replaces one problem — calculating a definite integral — with another: That of finding an *antiderivative* of a function (explained below). If the latter problem, which on the surface appears to be totally unrelated to the first, is solvable (it isn't always), then the solution provides an easy method for evaluating the integral.

By an *antiderivative* of a function f on an interval $[a, b]$ we mean another function, y, with the property that $g'(x) = f(x)$ for all x in $[a, b]$. For example, x^2 is an antiderivative of $2x$. So, for that matter, is $x^2 + 5$ or $x^2 + C$, where C is any constant (Figure 4-26).

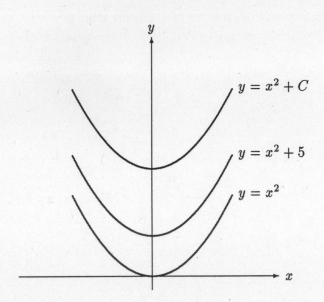

Figure 4-26: Antiderivatives of $2x$

Similarly, $\sin x + C$ is an antiderivative of $\cos x$ for any constant C. Thus, we see that antiderivatives are not unique, which is not surprising, since, geometrically, knowledge of the derivative of a function merely specifies the *slope* of its graph at each point, and the slope by itself does not determine the *position* of the curve, but only the *direction* in which it is moving. However, it can be shown that the graphs of all antiderivatives of a function are vertical shifts of one another (see Figure 4-26). In analytic terms, this means that if g is *some* antiderivative of f, then *any* other antiderivative of f must be of the form $g + C$, for some constant, C.

But what do antiderivatives have to do with integrals? Let's return to our examples of Section 4.3. In Example 4.1 we showed that $\int_0^1 x \, dx = 1/2$. Suppose we change the interval of integration from $[0, 1]$ to $[0, b]$. Rather than employing the same technique that we used in Example 4.1, we merely note that we are now finding the area of an isosceles right triangle with legs b. The area of a such a triangle is clearly $b^2/2$. Thus

$$\int_0^b x \, dx = \frac{b^2}{2}.$$

Turn next to the most general interval, $[a, b]$. We find here that

$$\int_a^b x \, dx = \frac{b^2}{2} - \frac{a^2}{2}.$$

(This integral is simply the area of one triangle minus the area of a second.)

Now, let's repeat these steps in connection with Example 4.2. By a computation similar to that done in the original problem (the details are in Solved Problem 4.10), we can first show that

$$\int_0^b x^2 \, dx = \frac{b^3}{3},$$

and then that

$$\int_a^b x^2 \, dx = \frac{b^3}{3} - \frac{a^3}{3}.$$

Can we make any sense of these results? Well, let's look for a connection between the *integrands*, x and x^2, and the outcomes of these calculations.

$$\frac{b^2}{2} - \frac{a^2}{2} \quad \text{arose from integrating} \quad x;$$

$$\frac{b^3}{3} - \frac{a^3}{3} \quad \text{arose from integrating} \quad x^2.$$

Now $x^2/2$ and $x^3/3$ are *antiderivatives* of x and x^2, respectively. So it appears from these examples that a simple method for *computing* a definite integral consists of finding an antiderivative of the integrand, and then subtracting its value at the lower limit of integration from its value at the upper limit. The Fundamental Theorem which follows confirms that our guess is correct.

Theorem 4.1 (The Fundamental Theorem of Calculus)
If g is any antiderivative of f, then

$$\int_a^b f(x) \, dx = g(b) - g(a) \qquad (4.17)$$

The Fundamental Theorem often makes the calculation of integrals quite trivial.

Example 4.3   *Compute $\int_0^1 (x^2 - 3x + 5) \, dx$.*

Solution: An antiderivative of $f(x) = x^2 - 3x + 5$ is $g(x) = x^3/3 - 3x^2/2 + 5x$. Hence, from (4.17),

$$\int_0^1 (x^2 - 3x + 5) \, dx = g(1) - g(0) = \frac{1}{3} - \frac{3}{2} + 5 = \frac{23}{6}.$$

Look how easy it is to evaluate this integral using the Fundamental Theorem, especially in comparison with our earlier computations in Example 4.1 (page 106) and Example 4.2 (page 108).

Example 4.4   *Compute $\int_0^1 x^k \, dx$.*

$g(x) = x^{k+1}/(k+1)$ is an antiderivative of x^k, so that

$$\int_0^1 x^k \, dx = g(1) - g(0) = \frac{1}{k+1}.$$

Example 4.5

$$\int_0^{\pi/2} \cos x \, dx = \sin \frac{\pi}{2} - \sin 0 = 1.$$

Further notation: Because the expression $g(b) - g(a)$ occurs so frequently in this subject, it is convenient to abbreviate it. We write

$$g(x)|_a^b = g(b) - g(a).$$

Thus, for example, if $g(x) = x^3/3$, then $g(1) - g(0)$ is written as

$$g(x)|_0^1 = \frac{x^3}{3}\bigg|_0^1 = \frac{1}{3} - \frac{0}{3} = \frac{1}{3}.$$

The advantage of this notation lies in the fact that we no longer have to give a 'name' to an antiderivative of a function. Previously, if we wanted to evaluate, say, $\int_0^1 x^2 \, dx$ by the Fundamental Theorem, then we had to 'say' the following:

- $g(x) = x^3/3$ is an antiderivative of x^2 ('naming' the antiderivative).

- Hence, $\int_0^1 x^2 \, dx = g(1) - g(0) = 1/3 - 0/3 = 1/3$.

With our new notation, we simply write

$$\int_0^1 x^2 \, dx = \frac{x^3}{3}\bigg|_0^1 = \frac{1}{3} - \frac{0}{3} = \frac{1}{3}.$$

Let's explore this new notation some more. The standard notation for the antiderivative of f is

$$\int f(x) \, dx.$$

But wait!

$$\int_a^b f(x) \, dx$$

is the notation for the definite integral. How can we tell them apart? The answer is easy (although the similarity of notation may still be troublesome): The definite integral has numbers at the top and bottom of the integral sign, $\int$. Thus, $\int_0^1 x^2 \, dx$ represents the *definite integral* of x^2 on the interval $[0, 1]$. It is a *number* (1/3 in this case). On the other hand, $\int x^2 \, dx$ represents an *antiderivative* of x^2. It is a *function* ($x^3/3$ is one such antiderivative, $x^3/3 + 10$ is another, and $x^3/3 + C$ is the most general one).

O.K., so we see the difference between $\int_a^b f(x) \, dx$ and $\int f(x) \, dx$. But isn't it absurd to represent two distinct concepts with symbols so similar that confusion is likely to result? The Fundamental Theorem of Calculus contains the answer: The two concepts are intimately linked. In fact, using the notation just introduced, we have, by the Fundamental Theorem,

$$\int_a^b f(x)\,dx = \left[\int f(x)\,dx\right]\Big|_a^b,$$

since $\int f(x)\,dx$ instructs us to find an antiderivative of f and the symbol $[\quad]\big|_a^b$ tells us to subtract the value of this antiderivative at a from its value at b. For example,

$$\int_0^1 x^2\,dx = \left[\int x^2\,dx\right]\Big|_0^1 = \left[\frac{x^3}{3}\right]\Big|_0^1 = \frac{1}{3} - \frac{0}{3} = \frac{1}{3};$$

$$\int_0^{\pi/2} \cos x\,dx = \left[\int \cos x\,dx\right]\Big|_0^{\pi/2} = \sin x\big|_0^{\pi/2} = 1 - 0 = 1.$$

(Note that we wrote $\int x^2\,dx = x^3/3$ and not $x^3/3 + C$, which is the most general antiderivative of x^2. The reason for this is that the Fundamental Theorem of Calculus allows us to use *any* antiderivative and we chose the simplest one.) We see that the Fundamental Theorem and the notation $\int f(x)\,dx$ for antiderivatives give us a super shorthand for the evaluation of definite integrals.

Exercise 4.5 *Evaluate the following definite integrals using the Fundamental Theorem:*
a. $\int_0^2 (x^2 + 3x - 4)\,dx$;   b. $\int_0^\pi (x + \cos x)\,dx$.

Because the use of antiderivatives so simplifies the computation of definite integrals, extensive techniques for the calculation of antiderivatives have been developed. You can find these methods in the chapter of your text entitled Techniques (or Methods) of Integration. Here, we present a short table of integrals which includes some of the most important functions.

Table of Integrals

1. $\displaystyle\int cg(x)\,dx = c\int g(x)\,dx$

2. $\displaystyle\int (g(x) \pm h(x))\,dx = \int g(x)\,dx \pm \int h(x)\,dx$

3. $\displaystyle\int x^n\,dx = \frac{x^{n+1}}{n+1} + C, \ \ n \neq 1$

4. $\displaystyle\int \frac{dx}{x} = \ln|x| + C$

5. $\displaystyle\int \sin x\,dx = -\cos x + C$

6. $\displaystyle\int \cos x \, dx = \sin x + C$

7. $\displaystyle\int \tan x \, dx = -\ln|\cos x| + C$

8. $\displaystyle\int \cot x \, dx = \ln|\sin x| + C$

9. $\displaystyle\int \sec x \, dx = \ln|\sec x + \tan x| + C$

10. $\displaystyle\int \csc x \, dx = -\ln|\csc x + \cot x| + C$

11. $\displaystyle\int \sec^2 x \, dx = \tan x + C$

12. $\displaystyle\int \csc^2 x \, dx = -\cot x + C$

13. $\displaystyle\int \sec x \tan x \, dx = \sec x + C$

14. $\displaystyle\int \csc x \cot x \, dx = -\csc x + C$

15. $\displaystyle\int e^x \, dx = e^x + C$

16. $\displaystyle\int a^x \, dx = \frac{a^x}{\ln a} + C$

17. $\displaystyle\int \frac{dx}{\sqrt{1 - x^2}} = \sin^{-1} x + C$

18. $\displaystyle\int \frac{dx}{1 + x^2} = \tan^{-1} x + C$

19. $\displaystyle\int \frac{dx}{x\sqrt{x^2 - 1}} = \sec^{-1} x + C$

20. $\displaystyle\int x e^x \, dx = x e^x - e^x + C$

21. $\displaystyle\int \ln x \, dx = x \ln x - x + C$

22. $\displaystyle\int x \sin x \, dx = \sin x - x \cos x + C$

23. $\displaystyle\int x \cos x \, dx = \cos x + x \sin x + C$

24. $\displaystyle\int e^x \sin x \, dx = \frac{e^x}{2}(\sin x - \cos x) + C$

25. $\displaystyle\int e^x \cos x \, dx = \frac{e^x}{2}(\cos x + \sin x) + C$

4.5 Applications of the Integral

There are numerous applications of the integral, in areas which include mathematics, all of the sciences, engineering, medicine, economics and other social sciences. In mathematics, we find areas of irregularly shaped regions, volumes of certain 3-dimensional solids, and lengths of curves by means of integrals. The integral arises in probability theory, and, in calculus, it is used to define new functions (for example, the natural logarithm). In physics, the integral allows us to compute liquid pressure in a tank, the flow of a fluid through a pipe (or blood vessel), and the center of gravity of a body. The integral also appears in population growth and finance. You will find many of these applications in your text, generally in a chapter entitled Applications of the Integral. In this book, we'll study a few of the applications, which are found in the next chapter.

Solved Problems

4.1 Suppose the interval $[0, 2]$ is subdivided by the partition $P = \{0, \frac{2}{3}, 1, \frac{3}{2}, \frac{7}{4}, 2\}$. What is $\|P\|$?

Solution: The mesh of the partition is the *largest* subinterval. In this case, $\|P\| = 2/3$.

4.2 Compute Riemann sums which approximate the integral $\int_1^2 (1/x) \, dx$ by partitioning the interval $[1, 2]$ into 4 equal subintervals, and taking c_i, $i = 1, 2, 3, 4$ to be

a. the right-hand endpoint of each subinterval

b. the left-hand endpoint of each subinterval

c. the midpoint of each subinterval.

Solution: In all of the cases $\Delta x = .25$, and the partition points are $x_0 = 1$, $x_1 = 1.25$, $x_2 = 1.5$, $x_3 = 1.75$ and $x_4 = 2$.

a. Here, $c_i = x_i$, $i = 1, 2, 3, 4$, so the sum becomes

$$\left[\frac{1}{1.25} + \frac{1}{1.5} + \frac{1}{1.75} + \frac{1}{2}\right].25 = .6345.$$

Because the integrand is decreasing, its smallest value occurs at the right-hand endpoint of each subinterval. Hence, this choice of c_i yields a lower sum for the integral.

b. In this case, $c_i = x_{i-1}, \quad i = 1, 2, 3, 4$. The Riemann sum is thus

$$\left[1 + \frac{1}{1.25} + \frac{1}{1.5} + \frac{1}{1.75}\right].25 = .7595.$$

This choice of c_i yields an upper sum, since the largest value of the integrand occurs at the left-hand endpoint of each subinterval.

c. In the final case, $c_1 = 9/8, \quad c_2 = 11/8, \quad c_3 = 13/8$ and $c_4 = 15/8$. The Riemann sum is

$$\left[\frac{8}{9} + \frac{8}{11} + \frac{8}{13} + \frac{8}{15}\right].25 = .6912.$$

The true value of the integral (to 4 decimal places) is .6931, so we see that choosing the midpoints of the subintervals gives the most accurate estimate in this case. In fact, the midpoints are generally a good choice, although there is no guarantee that they will always give the best result. We'll have more to say about this subject in the section on Numerical Integration in Chapter 6.

4.3 Express the following limits as definite integrals:

a. $\lim\limits_{||P|| \to 0} \sum\limits_{i=1}^{n} (c_i^4 + 3c_i^2 - 7c_i + 10)\Delta x_i$, where P is a partition of $[0, 1]$.

b. $\lim\limits_{||P|| \to 0} \sum\limits_{i=1}^{n} (\tan c_i - \cot c_i)\Delta x_i$, where P is a partition of $[\pi/6, \pi/4]$.

c. $\lim\limits_{||P|| \to 0} \sum\limits_{i=1}^{n} \sqrt{c_i + 1}(c_i - 4)\Delta x_i$, where P is a partition of $[1, 5]$.

Solution:

a. $\int_0^1 (x^4 + 3x^2 - 7x + 10)\, dx.$

b. $\int_{\pi/6}^{\pi/4} (\tan x - \cot x)\, dx.$

c. $\int_1^5 \sqrt{x + 1}(x - 4)\, dx.$

4.4 One of the nice things about the Fundamental Theorem of Calculus is that it does not matter how you come up with an antiderivative of the integrand. As long as the function you've produced is an antiderivative (which can be verified by differentiation), you may use it to evaluate the integral. This observation makes guessing a legitimate method of finding antiderivatives. (There are also systematic techniques in your text, but guessing should not be neglected.)

Verify that $x \ln x - x$ is an antiderivative of $\ln x$, and use this result to evaluate $\int_1^2 \ln x\, dx.$

Solution: Let $f(x) = x \ln x - x$. Then $f'(x) = \ln x + x \cdot (1/x) - 1 = \ln x$. Hence,

$$\int_1^2 \ln x \, dx = (x \ln x - x)\Big|_1^2 = (2 \ln 2 - 2) - (-1) = 2 \ln 2 - 1.$$

4.5 Verify that

$$f(x) = \frac{1}{2} \tan^{-1} \frac{x}{2}$$

is an antiderivative of $1/(4 + x^2)$. Use this result to evaluate

$$\int_0^2 \frac{dx}{4 + x^2}.$$

Solution: By the Chain Rule,

$$f'(x) = \left(\frac{1}{2}\right)\left(\frac{1}{1 + (x/2)^2}\right)\left(\frac{1}{2}\right) = \left(\frac{1}{2}\right)\left(\frac{4}{4 + x^2}\right)\left(\frac{1}{2}\right) = \frac{1}{4 + x^2}.$$

Hence,

$$\int_0^2 \frac{dx}{4 + x^2} = \frac{1}{2} \tan^{-1} \frac{x}{2}\Big|_0^2 = \frac{1}{2} \tan^{-1} 1 = \frac{1}{2}\frac{\pi}{4} = \frac{\pi}{8}.$$

4.6 Compute the following antiderivatives:

a. $\int (4xe^x - 3x^2 + \sqrt{x}) \, dx$ **b.** $\int (\ln x + 6 \sin x) \, dx$

c. $\int (x \sin x - x \cos x) \, dx$

Solution:

a. From numbers 1 and 2 of the Table of Integrals,

$$\int (4xe^x - 3x^2 + \sqrt{x}) \, dx = 4 \int xe^x \, dx - 3 \int x^2 \, dx + \int \sqrt{x} \, dx.$$

We now use numbers 3 and 20 to evaluate these three integrals.

$$4 \int xe^x \, dx - 3 \int x^2 \, dx + \int \sqrt{x} \, dx = 4xe^x - 4e^x - x^3 + \frac{2}{3}x^{3/2} + C.$$

b. We use numbers 1, 2, 5, and 21, and obtain

$$\int (\ln x + 6 \sin x) \, dx = \int \ln x \, dx + 6 \int \sin x \, dx = x \ln x - x - 6 \cos x + C.$$

c. We use 2, 22, and 23.

$$\int (x \sin x - x \cos x) \, dx = \sin x - x \cos x - \cos x - x \sin x + C.$$

In each of the above parts, verify the result by differentiating the outcome and seeing that it equals the integrand.

4.7 We've seen that antiderivatives are not unique: If g is an antiderivative of f, then so is $g + C$ for any constant C. However, if we know the value of the antiderivative at just *one* point, then we can determine it completely.

In each of the following, find an equation for a curve whose slope at x is $f(x)$ and which passes through the point (x_0, y_0).

a. $f(x) = 3x^2 + 5$, $(x_0, y_0) = (1, 10)$.

b. $f(x) = \sin x + x^3$, $(x_0, y_0) = (0, 1)$.

Solution:

a. The general antiderivative of f is $g(x) = x^3 + 5x + C$. Since the curve, $y = g(x)$, passes through the point $(1, 10)$, we must have $g(1) = 10$. Substituting, we find that $10 = g(1) = 1 + 5 + C$, so that $C = 4$. Hence, $y = x^3 + 5x + 4$ is the equation of the curve.

b. Here, $g(x) = -\cos x + x^4/4 + C$ is the general antiderivative of f, and we have to find C so that the curve $y = g(x)$ passes through $(0, 1)$. Substituting, we obtain $1 = g(0) = -1 + C$, and solving for C yields $C = 2$. Thus, $y = -\cos x + x^4/4 + 2$ is the equation.

4.8 Evaluate the following integrals:

a. $\displaystyle\int_1^2 \left(2x + \frac{1}{x}\right) dx$ **b.** $\displaystyle\int_4^9 \sqrt{x}\, dx$ **c.** $\displaystyle\int_0^1 e^{3x}\, dx$

d. $\displaystyle\int_0^{\pi/4} \cos 2x\, dx$ **e.** $\displaystyle\int_0^{1/2} \frac{dx}{\sqrt{1 - x^2}}$

Solution:

a.
$$\int_1^2 \left(2x + \frac{1}{x}\right) dx = (x^2 + \ln x)\Big|_1^2 = (4 + \ln 2) - 1 = 3 + \ln 2$$

b.
$$\int_4^9 \sqrt{x}\, dx = \frac{2}{3} x^{3/2}\Big|_4^9 = \frac{2}{3}(27 - 8) = \frac{38}{3}$$

c.
$$\int_0^1 e^{3x}\, dx = \frac{e^{3x}}{3}\Big|_0^1 = \frac{e^3 - 1}{3}$$

d.
$$\int_0^{\pi/4} \cos 2x\, dx = \frac{\sin 2x}{2}\Big|_0^{\pi/4} = \frac{1}{2}$$

e.

$$\int_0^{1/2} \frac{dx}{\sqrt{1-x^2}} = \sin^{-1} x \Big|_0^{1/2} = \sin^{-1}(1/2) - \sin^{-1}(0) = \frac{\pi}{6}.$$

4.9   Let $f(x) = \int_0^1 \sqrt{1+t^2}\, dt$. Find $f'(x)$.

Solution: $f'(x) = 0$! Since $f(x)$ is equal to a definite integral, which is a *number*, f is a *constant* function whose derivative is 0.

4.10   Let $A(t)$ be the area under the curve $f(x) = x^2$, $0 \le x \le t$ (Figure 4-27).

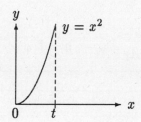

Figure 4-27: Area function

a.   Find $A(t)$ and $A'(t)$.

b.   Find the area under $y = x^2$, $a \le x \le b$.

Solution:

a.   We divide the interval $[0, t]$ into n subintervals of equal size $\Delta x = t/n$. The partition points will then be $0, t/n, 2t/n, \ldots, (n-1)t/n, t$. For c_i, we choose the righthand endpoint of each subinterval. The Riemann sum which approximates the integral is then

$$\left[\left(\frac{t}{n}\right)^2 + \left(\frac{2t}{n}\right)^2 + \left(\frac{3t}{n}\right)^2 + \cdots + \left(\frac{nt}{n}\right)^2 \right] \frac{t}{n} = \left(\frac{t}{n}\right)^3 \sum_{i=1}^n i^2.$$

Now, we saw in Example 4.2, page 108, that

$$\sum_{i=1}^n i^2 = \frac{n(n+1)(2n+1)}{6},$$

so that our Riemann sum is equal to

$$\left(\frac{t}{n}\right)^3 \left(\frac{n(n+1)(2n+1)}{6} \right) = t^3 \left[\frac{n(n+1)(2n+1)}{6n^3} \right].$$

As $n \to \infty$, the term in brackets $\to 1/3$. Hence, $A(t) = t^3/3$; so that $A'(t) = t^2 = f(t)$. Thus, the 'area function,' A, is an antiderivative of the integrand, f.

b. The area under the curve $y = x^2$, $a \leq x \leq b$ is equal to the area under the curve from 0 to b (which equals $A(b)$) minus the area under the curve from 0 to a (which equals $A(a)$). Thus, the area we are looking for is equal to

$$A(b) - A(a) = \frac{b^3}{3} - \frac{a^3}{3}.$$

4.11 In Figure 4-28, the areas of the indicated regions are as follows: $A_1 = 11$, $A_2 = 4$, $A_3 = 6$. Compute $\int_2^8 f(x)\,dx$.

Solution: Recall the connection between the integral and area. In computing

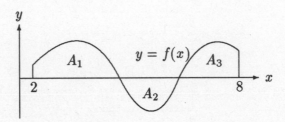

Figure 4-28: Area and the integral

$\int_2^8 f(x)\,dx$, the area of regions *above* the x-axis are counted as *positive*, while those *below* the axis are considered *negative*. Thus,

$$\int_2^8 f(x)\,dx = A_1 - A_2 + A_3 = 11 - 4 + 6 = 13.$$

4.12 Find the area between the curves $y = x^2$ and $y = x$ (Figure 4-29).

Solution: As we see in Figure 4-29, the two curves intersect at the points $x = 0$

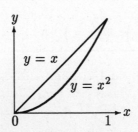

Figure 4-29: Area between two curves

and $x = 1$. Moreover, throughout this interval, the curve $y = x$ lies *above* $y = x^2$. Hence, the area between the two curves equals

$$\int_0^1 (x - x^2)\,dx = \left(\frac{x^2}{2} - \frac{x^3}{3} \right)\Big|_0^1 = \frac{1}{2} - \frac{1}{3} = \frac{1}{6}.$$

VELOCITY

When considering problems of motion, recall that three functions play an important role: The position function, s; the velocity function, v, which is the derivative of s; and the acceleration function, a, which is the derivative of v and, hence, the second derivative of s. One of the most common applications of these ideas is to freely falling bodies, that is, bodies that fall under the influence of gravity alone, and no other forces. This assumption, of course, is not really valid on earth, since it ignores air resistance. (It would be valid on the moon, which has no atmosphere.) But even on earth, the model that we'll use can serve as a first approximation, at least for certain falling bodies, such as balls, although not for others, such as feathers or parachutists. More sophisticated models are available to handle the excluded cases, but the basic principles can be learned from the simpler approach.

A freely falling body has *constant acceleration*, $a(t) = -32$ ft/sec^2. (The sign is negative because we adopt the obvious convention that *up* is the *positive* direction and *down* is *negative*.) Since $a(t) = v'(t)$, we have

$$v(t) = \int a(t)\,dt = -32t + v_0,$$

where the constant of integration is denoted by v_0, the *initial velocity*. Similarly,

$$s(t) = \int v(t)\,dt = -16t^2 + v_0 t + s_0,$$

with s_0 representing the *initial position* or height of the body. The three equations,

$$s(t) = \quad -16t^2 + v_0 t + s_0$$
$$v(t) = \quad -32t + v_0$$
$$a(t) = \quad -32$$

give us complete information about the motion of the body.

4.13 Suppose a ball is *dropped* from a window which is 100 feet above the ground.

a. How long does it take for the ball to reach the ground?

b. What is the velocity of the ball at the time it hits the ground?

Solution:

a. The conditions in the problem tell us that $v_0 = 0$ (the ball was *dropped*, not *thrown* down) and $s_0 = 100$ (the height from which it was released). Thus, $s(t) = -16t^2 + 100$. At ground level, $s = 0$, so we set $s(t) = 0$ and solve for t:

$$16t^2 + 100 \quad - \quad 0$$
$$16t^2 \quad = \quad 100$$

$$t^2 \ = \ 100/16$$

$$t \ = \ 2.5$$

Thus, it takes 2.5 seconds for the ball to reach the ground.

b. $v(t) = -32t$, so that the velocity at time $t = 2.5$ is $-32(2.5) = -80$ feet per second (which is equivalent to 55 miles per hour — so look out for falling balls!).

4.14 You are trying to throw a ball to a friend who is at a window 25 feet above where you stand.

a. You throw the ball with an initial velocity of $v_0 = 32$ ft/sec. Assuming that your aim is accurate, will the ball reach your friend?

b. What is the smallest possible value of v_0 which will accomplish your purpose?

Solution:

a. $v(t)$ is equal to $-32t + 32$ ft/sec. The ball will reach its maximum height when $v(t) = 0$, which occurs at $t = 1$. Now $s(t) = -16t^2 + 32t$, so the ball reaches a height of just 16 feet, and falls short of your friend.

b. In this part, v_0 is the unknown. Here, $v(t) = -32t + v_0$ and $s(t) = -16t^2 + v_0 t$. As in the previous part, the maximum height occurs when $v(t) = 0$, which is at $t = v_0/32$. Substituting this value of t into the expression for $s(t)$, we obtain

$$s_{\max} = -16 \left(\frac{v_0}{32} \right)^2 + \frac{v_0^2}{32} = \frac{v_0^2}{64}.$$

Now we want the maximum height to be at least 25 feet, so we set $s_{\max} = 25$, or

$$\frac{v_0^2}{64} = 25.$$

Solving, we obtain $v_0 = 40$ ft/sec. as the miminum initial velocity.

4.15 Suppose now that the ball is *thrown* directly up from the window with an initial velocity of 60 ft/sec. Answer parts (a) and (b) Solved Problem 4.13 as well as these:

c. At what time does the ball reach its maximum height?

d. What is its maximum height?

Solution: The equations of motion in this case are

$$\begin{aligned} s(t) &= \quad -16t^2 + 60t + 100 \\ v(t) &= \quad -32t + 60 \\ a(t) &= \quad -32 \end{aligned}$$

a. Set $s(t) = 0$ and solve the equation $-16t^2 + 60t + 100 = 0$, obtaining (from the quadratic formula) $t = 5$ seconds and $t = -5/4$ seconds. The second solution is a valid one for the quadratic equation, but *not for the physical situation it represents*, since t must be positive. ($t = 0$ corresponds to the moment the ball is released.) Hence, $t = 5$ seconds is the solution we're looking for.

b. The velocity at time $t = 5$ is $v(5) = -160 + 60 = -100$ ft/sec.

c. When the ball reaches its maximum height, it stops rising and begins to fall. Hence, the velocity changes from positive (rising) to negative (falling), which means that $v(t) = 0$ at that time. So set $v = 0$ and solve $v(t) = -32t + 60 = 0$, yielding $t = 1.875$ seconds.

d. The maximum height is equal to $s(1.875) = 156.25$ feet.

4.16 Gravity on the moon is approximately $1/6$ of that on earth. Moreover, there is no atmosphere, so there is no air resistance. Hence, the form of the model of a freely falling body that we developed is valid on the moon, with just some changes in the coefficients due to the difference in the gravitational constant. Specifically, the acceleration of a freely falling body is $a = -5.2$ ft/sec^2, so that integration gives us $v = -5.2t + v_0$ and $s = -2.6t^2 + v_0 t + s_0$ as the equations of motion.

Suppose that one of the astronauts had dropped a rock from the entrance to the lunar module, which is 20 feet above the moon's surface. How long would it have taken for the rock to reach the ground?

Solution: In this problem, $v_0 = 0$ and $s_0 = 20$, so that $s = -2.6t^2 + 20$. To find the time at which the rock hits the ground, we set this last equation equal to 0 and solve. We obtain $t = \sqrt{\frac{20}{2.6}} = 2.8$ seconds.

4.17 A spider is climbing up a 9-foot tall tree, in order to eat a fly trapped in a web at the very top of the tree. At 1 P.M. the spider is at ground level and its velocity at time t hours is $v(t) = 18/t^3$ feet per hour.

a. How far does the spider go between 1 P.M. and 3 P.M.?

b. Will the spider ever reach the top of the tree?

Solution:

a. The distance traveled between 1 P.M. and 3 P.M. equals

$$\int_1^3 v(t)\, dt = \int_1^3 \frac{18}{t^3}\, dt = -\frac{9}{t^2}\Big|_1^3 = 8 \text{ feet.}$$

b. No, the spider never makes it to the top. For, let b be *any* time. Let's calculate the distance the spider travels between $t = 1$ and $t = b$, which equals

$$\int_1^b v(t)\,dt = \int_1^b \frac{18}{t^3}\,dt = -\frac{9}{t^2}\Big|_1^b = 9 - \frac{9}{b^2}\text{ feet.}$$

This quantity is always less than 9, since $9/b^2$ is positive. Hence, the spider doesn't reach the top.

4.18 Find the *total distance* traveled by a body moving in a straight line with velocity $v(t) = t^2 - 3t + 2,\ \ 0 \le t \le 2$.

Solution: Recall that the distance traveled equals the integral of the velocity function only when the velocity is positive throughout the time interval. Otherwise, the integral represents the *displacement* of the body from its initial position. In this case, the velocity is *not* always positive. In fact, $v(t) = (t-1)(t-2)$, so that v is positive for $0 \le t \le 1$ and negative for $1 \le t \le 2$. So to find the distance traveled we have to compute the integral of the *absolute value* of the velocity function:

$$
\begin{aligned}
\text{distance} &= \int_0^2 |v(t)|\,dt \\[2mm]
&= \int_0^1 (t^2 - 3t + 2)\,dt + \int_1^2 -(t^2 - 3t + 2)\,dt \\[2mm]
&= \left(\frac{t^3}{3} - \frac{3t^2}{2} + 2t\right)\Big|_0^1 - \left(\frac{t^3}{3} - \frac{3t^2}{2} + 2t\right)\Big|_1^2 \\[2mm]
&= \frac{5}{6} + \frac{1}{6} = 1.
\end{aligned}
$$

WORK

4.19 Two unlike charges attract each other with a force $1/x^2$, where x is the distance between the charges. One charge is fixed at $(1, 0)$ and the second is moved along the x-axis from $(2, 0)$ to $(4, 0)$. How much work is done?

Solution: The work done is the integral of the force function. Thus,

$$W = \int_2^4 \frac{1}{x^2}\,dx = -\frac{1}{x}\Big|_2^4 = -\frac{1}{4} - \left(-\frac{1}{2}\right) = \frac{1}{4}.$$

4.20 A force is needed to stretch (or compress) a spring. Many springs satisfy Hooke's Law, which states that the force required is proportional to the amount the spring is stretched from its natural position. The constant of proportionality depends upon the material used in constructing the spring and in how tightly it is coiled. Suppose a spring has a natural length of 1 foot, and the force required to stretch it x feet is $24x$ pounds. Find the work done in stretching the spring to a length of 1.5 feet.

Solution: Since $F(x) = 24x$, the work done is given by

$$W = \int_0^{.5} 24x \, dx = 12x^2 \Big|_0^{.5} = 3 \text{ foot-pounds.}$$

Supplementary Problems

4.21 Is $P = \{1.1, 1.3, 1.5, 1.7, 1.9\}$ a partition of $[1,2]$?

4.22 Find the mesh of the partition $P = \{0, 1/3, 1/2, 3/4, 1\}$ of the interval $[0,1]$. How many subintervals are there? Write out the values of Δx_i.

4.23 Evaluate the following integrals. (You may use the table of integrals found on page 113.)

    a. $\int_0^2 \left(4x^3 - x^2 + 6x + 2\right) dx$ **b.** $\int_1^e \frac{4dx}{x}$

    c. $\int_1^4 \left(\sqrt{x} + \frac{1}{\sqrt{x}}\right) dx$ **d.** $\int_{\sqrt{2}}^2 \frac{dx}{x\sqrt{x^2 - 1}}$

4.24 Find the area under the following curves:

    a. $y = \sin x, \ 0 \le x \le \pi$ **b.** $y = \tan x, \ 0 \le x \le \pi/4$

    c. $y = e^x, \ 0 \le x \le 2$

4.25 Find the area between the curves $y = f(x)$ and $y = g(x)$. (**Hint:** First find the points of intersection of the two curves by setting $f(x) = g(x)$. The solutions of this equation give the limits of integration.)

    a. $f(x) = x^2 - 1$ and $g(x) = -x^2 + 3$.

    b. $f(x) = 2^x$ and $g(x) = x + 1$. (**Hint:** One of the points of intersection is when $x = 1$. Find the other one by inspection.)

4.26 Compute the following antiderivatives:

    a. $\int (\csc^2 x + 10^x) \, dx$ **b.** $\int (\sec x - 4e^x \cos x) \, dx$

4.27 A batter hits a ball directly up in the air. At the moment of impact, the ball is 4 feet above the ground and the velocity imparted to the ball by the bat is 64 ft/sec.

    a.   How long does it take the ball to reach its maximum height?

    b.   What is its maximum height?

    c.   The ball falls to the ground untouched. How long was the ball in the air?

4.28   The force required to stretch a spring x feet from its natural length is $F(x) = 10x$. Find the work done in stretching the spring 1 foot.

Answers to Supplementary Problems

4.21   No, since the endpoints, 1 and 2, are not included in P.

4.22   Mesh $= 1/3$;   4 subintervals; $\Delta x_1 = 1/3$; $\Delta x_2 = 1/6$; $\Delta x_3 = 1/4$; $\Delta x_4 = 1/4$.

4.23   **a.** $89/3$ **b.** 4 **c.** $20/3$ **d.** $\pi/12$

4.24   **a.** 2 **b.** $\ln\sqrt{2}$ **c.** $e^2 - 1$

4.25   **a.** $\dfrac{16\sqrt{2}}{3}$ **b.** $\dfrac{3}{2} - \dfrac{1}{\ln 2}$

4.26   **a.** $-\cot x + \dfrac{10^x}{\ln 10} + C$ **b.** $\ln|\sec x + \tan x| - 2e^x(\cos x + \sin x)$

4.27   **a.** 2 seconds **b.** 68 feet **c.** $(4 + \sqrt{17})/2$ seconds

4.28   5 foot-pounds.

Answers to Exercises

4.1   (Page 93) **a.** 204 **b.** $5/6$

4.2   (Page 93) **a.** $\displaystyle\sum_{i=1}^{10} 2i$ **b.** $\displaystyle\sum_{i=1}^{n} a_i b_i$

4.3   (Page 105) **a.** $\displaystyle\sum_{i=1}^{n}(c_i^2 + 3c_i + 1)\Delta x_i$ **b.** $\displaystyle\sum_{i=1}^{n}(c_i + \cos c_i)\Delta x_i$

4.4 (Page 106) **a.** $\displaystyle\int_0^1 (x^4 + 3x^2 - 2)\,dx$ **b.** $\displaystyle\int_0^1 \frac{x^2 + 1}{x^2 + 2x + 4}$

4.5 (Page 113) **a.** $2/3$ **b.** $\pi^2/2$

Chapter 5

Applications of the Integral

Many of the applications of the integral provide excellent illustrations of our organizing principle: Approximation—Refinement—Limit ($\mathcal{A}$—$\mathcal{R}$—$\mathcal{L}$). For example, we will use our knowledge of the length of a line segment in order to find the length of a curve, and the volume of a cylinder to determine volumes of much more general figures. As mentioned at the end of the last chapter, applications in geometry, physics, and other scientific areas abound, and you were referred there to your text for the details. Here, we elaborate on two of the applications, which emphasize the $\mathcal{A}$—$\mathcal{R}$—$\mathcal{L}$ theme.

5.1   Arc Length

What We Know: The length of a *straight line*.

What We Want To Know: The length of an arbitrary *curve*.

How We Do It: Approximate the *curve* with polygonal (broken) *lines*.

At the present time, our knowledge of length is restricted to straight lines. Of course, we are also familiar with the formula $C = 2\pi R$ for the circumference of a circle, but in most high school geometry courses this formula is merely presented and perhaps justified in an informal way, but not derived rigorously. We can, however, use trigonometry and the notion of limit to obtain this formula, and we begin our discussion of arc length with a derivation of this result.

Consider a regular polygon of n sides inscribed in a circle of radius R (Figure 5-1). (A regular polygon is one in which all *sides* and all *angles* are equal.) The sum of the lengths of the sides of this polygon is called the *perimeter*, P_n. P_n, which we now calculate, is an *approximation* to the circumference of the circle. If we connect the center of the circle, O, to each of the vertices of the polygon, we obtain n *congruent* triangles. (The triangles are congruent because all sides of the polygon are equal.) Now, the sum of the angles around the center, O, is 2π radians (or 360 degrees— recall, however, that in calculus we always employ radian measure). Since there are n

$\mathcal{A}$

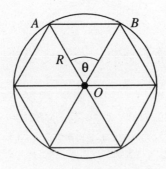

Figure 5-1: Approximating a circle with a polygon

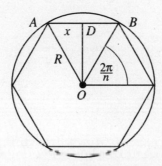

Figure 5-2: Calculating the perimeter P_n

equal central angles, the measure of each one of them is equal to $2\pi/n$. Now draw a perpendicular from the center O to one of the sides of the polygon (Figure 5-2). Let x be the length of the line segment AD. Then, since $\sin(\angle DOA) = x/R$, we have $x = R\sin(\angle DOA) = R\sin(\pi/n)$. Now x is half of the length of side AB, so that the length of one side of the polygon is $2R\sin(\pi/n)$. The perimeter of the polygon is found by multiplying this by n (the number of sides), thereby obtaining

$$P_n = 2nR\sin(\pi/n). \tag{5.1}$$

So we have found an exact formula for our *approximation* to the circumference, C. How do we *refine* or improve the approximation? Clearly, by using a regular polygon with more sides, which more closely resembles the circle (Figure 5-3). But regardless of the number of sides, n, equation (5.1) is valid for the perimeter. To obtain the *exact* value of C, we now pass to the *limit*; that is, we let the number of sides tend to infinity. Hence,

$$C = \lim_{n\to\infty} P_n = \lim_{n\to\infty} 2nR\sin(\pi/n). \tag{5.2}$$

But what is this limit? Recall, from the section in your text on trigonometric limits,

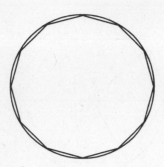

Figure 5-3: Refined approximation

that

$$\lim_{t \to 0} \frac{\sin t}{t} = 1. \tag{5.3}$$

In order to use (5.3), rewrite (5.2) as follows:

$$C = \lim_{n \to \infty} \frac{2R \sin(\pi/n)}{1/n} = \lim_{n \to \infty} \frac{2\pi R \sin(\pi/n)}{\pi/n}. \tag{5.4}$$

If we let $t = \pi/n$, then $n \to \infty$ causes t to approach 0. Hence,

$$C = \lim_{t \to 0} \frac{2\pi R \sin t}{t} = 2\pi R \lim_{t \to 0} \frac{\sin t}{t} = 2\pi R,$$

which recaptures the familiar formula.

Now, you may ask, where is the integral in all of this? In fact, it is not present, so perhaps you feel that the circle is not a good example of the methods we wish to illustrate. Nevertheless, the key feature of Approximation— Refinement—Limit appears here, as does the *type* of approximation that we will use to solve the general problem, namely, the use of a *polygon* to approximate a *curve*. The integral does not occur here only because the symmetry of the circle allows us to find the perimeter of the polygon by *multiplication* rather than by *addition*. Recall, that after finding the length of $AB = 2R \sin(\pi/n)$, we used the fact that the polygon is *regular* to find the perimeter by multiplying $2R \sin(\pi/n)$ by n. For curves less symmetric than a circle, we cannot avoid *summations* of the lengths of the individual line segments, and *summations of this type often lead to integrals.*

We now turn to the general case of arc length. Consider a curve which is the graph of a continuous function, $y = f(x)$, $a \le x \le b$ (Figure 5-4). We wish to find its length, L, and begin by fitting a polygonal (broken) line to the curve (Figure 5-5). This is hopefully a useful approach for two reasons: First, this method worked well for the circle, as we saw earlier. Second, since, with the exception of the circle, our knowledge of lengths is confined to straight lines, we have nothing else to rely upon. But how do we choose the particular polygonal line? In a fashion reminiscent of the construction of the

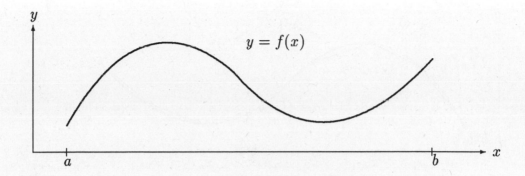

Figure 5-4: The curve whose length we're seeking

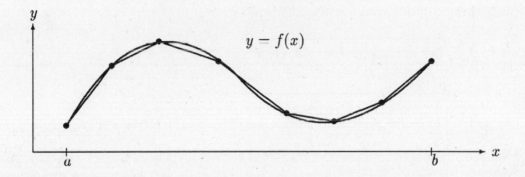

Figure 5-5: Approximation by a polygonal line

integral (in fact, we are going to construct a *certain* integral), we begin by partitioning $[a, b]$ into a number of subintervals by introducing a set of points

$$P = a = x_0 < x_1 < x_2 < \ldots < x_{n-1} < x_n = b.$$

The partition points $x_0, x_1, x_2, \ldots, x_n$, determine points on the curve (x_0, y_0), (x_1, y_1), $(x_2, y_2), \ldots, (x_n, y_n)$, where $y_0 = f(x_0)$, $y_1 = f(x_1)$, $\ldots$, $y_n = f(x_n)$. We now construct the line segments joining each consecutive pair of points, (x_0, y_0) with (x_1, y_1), (x_1, y_1) with (x_2, y_2), and so forth (Figure 5-6). The next step is to compute the length of this polygonal path, which serves as an *approximation* to L, the length of the curve. To do this, we use the formula for the distance between two points. Thus, the length of the first segment is $\sqrt{(x_1 - x_0)^2 + (y_1 - y_0)^2}$, the second segment has length $\sqrt{(x_2 - x_1)^2 + (y_2 - y_1)^2}$, and so forth. Hence, the total length of the polygon is

$$L_n = \sum_{i=1}^{n} \sqrt{(x_i - x_{i-1})^2 + (y_i - y_{i-1})^2}. \tag{5.5}$$

We are now going to manipulate expression (5.5) in order to obtain a Riemann sum (page 104), the limit of which will give us an integral formula for the arc length. To

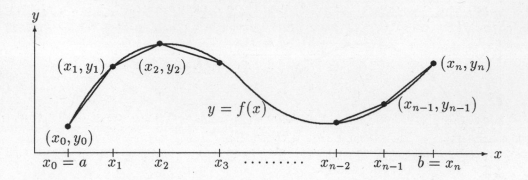

Figure 5-6: Initial approximation

accomplish this, we go through some algebraic steps.

$$L_n = \sum_{i=1}^{n} \sqrt{(x_i - x_{i-1})^2 + (y_i - y_{i-1})^2}$$

$$= \sum_{i=1}^{n} \sqrt{\left[1 + \left(\frac{y_i - y_{i-1}}{x_i - x_{i-1}}\right)^2\right](x_i - x_{i-1})^2} \qquad \textit{(factoring out } (x_i - x_{i-1})^2\textit{))}$$

$$= \sum_{i=1}^{n} \sqrt{1 + \left(\frac{y_i - y_{i-1}}{x_i - x_{i-1}}\right)^2}(x_i - x_{i-1}) \qquad (\sqrt{(x_i - x_{i-1})^2} = (x_i - x_{i-1}))$$

$$= \sum_{i=1}^{n} \sqrt{1 + \left(\frac{f(x_i) - f(x_{i-1})}{x_i - x_{i-1}}\right)^2}(x_i - x_{i-1}) \qquad \textit{(since } y_i = f(x_i)\textit{)}$$

$$= \sum_{i=1}^{n} \sqrt{1 + \left(\frac{f(x_i) - f(x_{i-1})}{x_i - x_{i-1}}\right)^2}\Delta x_i. \qquad \textit{(since } x_i - x_{i-1} = \Delta x_i\textit{)}$$

Summarizing, we have

$$L_n = \sum_{i=1}^{n} \sqrt{1 + \left(\frac{f(x_i) - f(x_{i-1})}{x_i - x_{i-1}}\right)^2}\Delta x_i. \qquad (5.6)$$

We are now close to a Riemann sum. The $\sum_{i=1}^{n}$ is present, as is Δx_i. But recall that the general form of a Riemann sum (for a function g) is

$$\sum_{i=1}^{n} g(c_i)\Delta x_i,$$

and c_i is *absent* from (5.6). We can, however, incorporate c_i into (5.6) by using the Mean Value Theorem of differential calculus. (**Recall**: If f is differentiable on an interval (u, v) and continuous on $[u, v]$, then there exists a point $c \in (u, v)$ such that $(f(v) - f(u))/(v - u) = f'(c)$.) Here we set $u = x_{i-1}$, $v = x_i$ and $c = c_i$ to obtain

$$L_n = \sum_{i=1}^{n} \sqrt{1 + (f'(c_i))^2} \Delta x_i, \tag{5.7}$$

and (5.7) *is* a Riemann sum for the function $\sqrt{1 + (f')^2}$. (And you thought that you'd never see any use for the Mean Value Theorem!)

Let's stop for a moment, before the details of the calculation swamp us. We have shown that:

1. A partition, P, of the interval $[a, b]$ gives rise to a polygonal line which approximates the curve $y = f(x)$;

2. The length, L_n, of this polygonal line is an approximation to the length, L, of the curve;

3. L_n can be expressed as

$$L_n = \sum_{i=1}^{n} \sqrt{1 + (f'(c_i))^2} \Delta x_i,$$

which is a Riemann sum for $\sqrt{1 + (f')^2}$.

To improve or *refine* the approximation, we take a finer partition of $[a, b]$ (Figure 5-7). But *each* such partition defines a polygonal line whose total length is given by the

$\mathcal{R}$

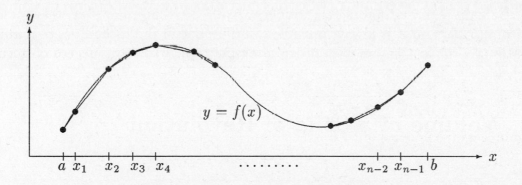

Figure 5-7: Refined approximation

Riemann sum (5.7). We now pass to the *limit*, proceeding on two fronts simultaneously. On the one hand, successive refinements of the interval produce polygonal lines, the sum of whose lengths, L_n, in a physical sense, more closely approximates the length of the curve, L. Hence, we can expect that

$\mathcal{L}$

$$L = \lim_{\|P\| \to 0} L_n, \tag{5.8}$$

where $||P|| = \max(x_i - x_{i-1})$, $i = 1, 2, \ldots, n$, is the mesh of the partition, P. On the other hand, we know that if the function $\sqrt{1 + (f')^2}$ is integrable, then as $||P|| \to 0$, the Riemann sums in (5.7) tend to the *integral*; that is,

$$\lim_{||P|| \to 0} \sum_{i=1}^{n} \sqrt{1 + (f'(c_i))^2} \Delta x_i = \int_a^b \sqrt{1 + (f'(x))^2}\, dx. \tag{5.9}$$

Combining (5.8) and (5.9) we arrive at the promised formula for the length of the curve, $y = f(x)$, $a \leq x \leq b$,

$$L = \int_a^b \sqrt{1 + (f'(x))^2}\, dx. \tag{5.10}$$

Example 5.1   *Find the length of the curve $y = (2/3)x^{3/2}$, $0 \leq x \leq 15$.*

Solution: Letting $f(x) = (2/3)x^{3/2}$, we have $f'(x) = x^{1/2}$. Hence, from (5.10),

$$\begin{aligned} L &= \int_0^{15} \sqrt{1 + (f'(x))^2}\, dx = \int_0^{15} \sqrt{1 + x}\, dx \\ &= \frac{2}{3}(1 + x)^{3/2} \Big|_0^{15} = \frac{2}{3}(64 - 1) = 42. \end{aligned}$$

Remark 5.1   Because of the form of the integrand in (5.10), it is difficult to produce many examples in which the integral can be evaluated using the Fundamental Theorem of Calculus (i.e., by finding an antiderivative). This fact, however, takes nothing away from (5.10) and it is always possible to utilize one of the techniques of numerical integration found in Chapter 6 in order to evaluate (5.10) to any degree of accuracy desired.

5.2   Volume of a Solid of Revolution

What We Know: The volume of a *cylinder*.

What We Want To Know: The volume of an arbitrary *solid of revolution*.

How We Do It: We approximate the *solid* with a collection of *cylinders*.

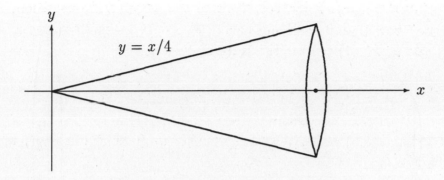

Figure 5-8: A cone is a solid of revolution

Certain solids can be obtained by rotating a figure in the plane about an axis. For example, if we rotate the region between the x-axis and line $y = x/4$, $0 \leq x \leq 4$ about the x-axis, we obtain a *cone* (Figure 5-8). Similarly, rotating the region between the x-axis and the semi-circle, $y = \sqrt{1 - x^2}$, $-1 \leq x \leq 1$, about the x-axis, generates a *sphere*. In this section, we will develop a method for calculating the volume of such *solids of revolution*.

Let f be a positive function in $[a, b]$, and let T be the region bounded above by the curve $y = f(x)$, below by the x-axis, and left and right by the lines $x = a$ and $x = b$, respectively (Figure 5-9). Revolve T about the x-axis, obtaining a body S which

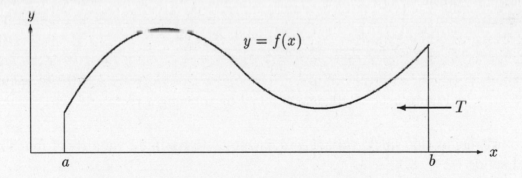

Figure 5-9: The region to be rotated

is called a solid of revolution (Figure 5-10). We wish to calculate its volume, V, and begin by considering the special case of a constant function, $f(x) = c$. In this case, the region T is simply a rectangle and the solid generated by its revolution about the x-axis is a *cylinder* (Figure 5-11).

The height of this cylinder is $(b - a)$ and the radius of its base is c. Now the volume of a cylinder of height h, and radius (of its base) r, is given by $\pi r^2 h$. So the volume of the cylinder in Figure 5-11 is $\pi c^2(b - a)$, or $\pi(f(x))^2(b - a)$, since $f(x) = c$. We will see

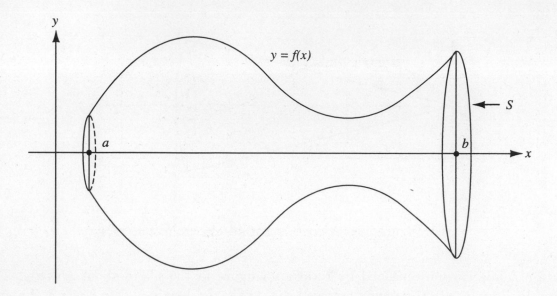

Figure 5-10: The solid of revolution

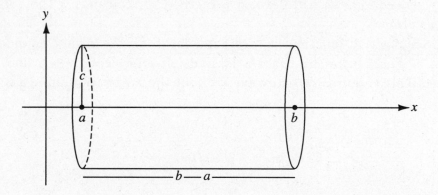

Figure 5-11: Cylinder obtained by rotating a rectangle

that knowledge of this single volume will enable us to compute volumes of much more general solids.

Our program for solving the general problem is as follows:

- We approximate the region T with a set of rectangles, obtained by partitioning the interval $[a, b]$ in the usual way, selecting in each subinterval $[x_{i-1}, x_i]$, $i = 1, 2, \ldots, n$, a point c_i, and constructing a rectangle of height $f(c_i)$ (Figure 5-12).

- We revolve each of these rectangles about the x-axis, thereby generating a collection of cylinders. The height of the ith cylinder is $(x_i - x_{i-1})$ and the radius of its base is $f(c_i)$ (Figure 5-13). (**Caution:** Notice in Figures 5-12 and 5-13 that the *height* of the *rectangle*, $f(c_i)$, is the *radius* of the base of the *cylinder*, while the *base* of the *rectangle*, $(x_i - x_{i-1})$, is the *height* of the *cylinder*.)

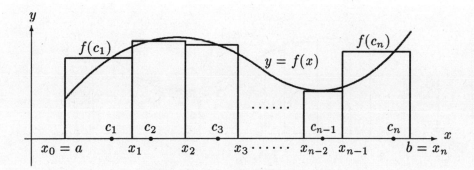

Figure 5-12: Approximating the region with rectangles

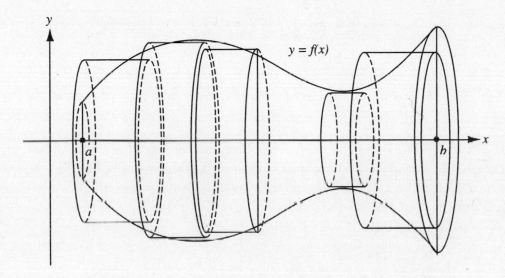

Figure 5-13: Approximating the solid with cylinders

- The sum of the volumes of these cylinders, which serves as an approximation to the volume, V, of the solid, is given by the following *Riemann sum* for the function $\pi(f)^2$,

$$\sum_{i=1}^{n} \pi(f(c_i))^2 (x_i - x_{i-1}). \tag{5.11}$$

- We now use the same reasoning as in our study of arc length. We *refine* the approximation by taking a finer partition of $[a, b]$, resulting in more rectangles (Figure 5-14). As the number of rectangles increases, the approximation improves. To obtain the exact volume we pass to the *limit*. Once again, two things happen to the Riemann sums in (5.11): Physically, they tend to V, while mathematically

$\mathcal{A}$

$\mathcal{R}$

$\mathcal{L}$

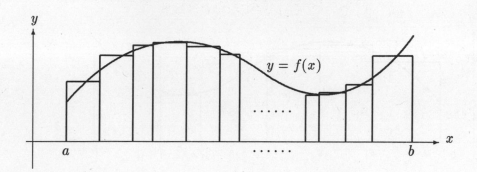

Figure 5-14: Refined approximation

they tend to the integral. In other words,

$$V = \lim_{\|P\| \to 0} \sum_{i=1}^{n} \pi(f(c_i))^2 \Delta x_i = \int_a^b \pi(f(x))^2 \, dx. \tag{5.12}$$

Example 5.2 *Let R be the region between the curve $y = \sin x$ and the x-axis, for $0 \le x \le \pi$. Find the volume of the solid generated when R is rotated around the x-axis.*

Solution: By (5.12), $V = \pi \int_0^\pi \sin^2 x \, dx$. In order to evaluate this integral, we employ a trigonometric identity, which enables us to find an antiderivative of $\sin^2 x$. (This is a standard method for the computation of antiderivatives of powers of sines and cosines, which you'll find in your text in the chapter on Techniques of Integration.) Specifically, $\sin^2 x = (1 - \cos 2x)/2$. Hence,

$$
\begin{aligned}
V &= \pi \int_0^\pi \sin^2 x \, dx \\[2mm]
&= \frac{\pi}{2} \int_0^\pi (1 - \cos 2x) \, dx \quad &\text{\textit{(substitution)}} \\[2mm]
&= \frac{\pi}{2}\left(x - \frac{\sin 2x}{2}\right)\Big|_0^\pi \quad &\text{\textit{(Fundamental Theorem)}} \\[2mm]
&= \frac{\pi}{2}(\pi - 0) \quad &\text{\textit{(substitution)}} \\[2mm]
&= \frac{\pi^2}{2}.
\end{aligned}
$$

Solved Problems

ARC LENGTH

5.1 Find the length of the curve

$$y = \frac{1}{3}(x^2 - 2)^{3/2}$$

from $x = 3$ to $x = 6$.

Solution: If we let $f(x) = (x^2 - 2)^{3/2}/3$, then $f'(x) = x\sqrt{x^2 - 2}$, so that

$$\sqrt{1 + (f'(x))^2} = \sqrt{1 + x^2(x^2 - 2)} = \sqrt{x^4 - 2x^2 + 1} = \sqrt{(x^2 - 1)^2} = x^2 - 1.$$

Therefore, the length of the curve is equal to

$$\int_3^6 \sqrt{1 + (f'(x))^2}\, dx = \int_3^6 (x^2 - 1)\, dx = \left(\frac{x^3}{3} - x\right)\Big|_3^6 = 60.$$

5.2 Find the length of the curve

$$y = \frac{x^3}{2} + \frac{1}{6x}$$

from $x = 2$ to $x = 4$.

Solution: Letting $f(x) = x^3/2 + 1/(6x)$, we obtain $f'(x) = (3x^2)/2 - 1/(6x^2)$. Hence,

$$\sqrt{1 + (f'(x))^2} = \sqrt{1 + \frac{9x^4}{4} - \frac{1}{2} + \frac{1}{36x^4}} = \sqrt{\frac{9x^4}{4} + \frac{1}{2} + \frac{1}{36x^4}}$$

$$= \sqrt{\left(\frac{3x^2}{2} + \frac{1}{6x^2}\right)^2} = \left(\frac{3x^2}{2} + \frac{1}{6x^2}\right).$$

Thus, the length of the curve equals

$$\int_2^4 \sqrt{1 + (f'(x))^2}\, dx = \int_2^4 \left(\frac{3x^2}{2} + \frac{1}{6x^2}\right) dx = \left(\frac{x^3}{2} - \frac{1}{6x}\right)\Big|_2^4 = \frac{673}{24}.$$

5.3 In general, the approximation to an integral using Riemann sums can be either larger or smaller than the integral. Show, however, that in the computation of the arc length of a curve, the approximations are *never* larger than the actual length.

Solution: Consider the portion of the curve between successive partition points, (x_{i-1}, y_{i-1}) and (x_i, y_i). The approximation to the length of this piece consists of joining these points by a *straight line*. Since the shortest distance between two points is the straight line joining them, the approximation is not larger than the actual length. Now just add up the pieces.

5.4   When we approximated velocity, work, area, and volume, we always used rectangles. In other words, we replaced a small segment of the given curve by a horizontal line. Why don't we use this method of approximation in the case of arc length?

Solution: Suppose we want to find the length of $y = f(x)$, $a \le x \le b$. If we use horizontal lines in each segment, then *every* approximation will have length $b - a$, which is clearly the wrong answer unless f is a constant function.

VOLUMES OF SOLIDS OF REVOLUTION

5.5   Find the volume of the solid obtained by rotating the region bounded by $f(x) = \sqrt{x}$, $1 \le x \le 7$, around the x-axis.

Solution: The volume is equal to

$$\int_1^7 \pi (f(x))^2 \, dx = \int_1^7 \pi x \, dx = \left. \frac{\pi x^2}{2} \right|_1^7 = 24\pi.$$

5.6   A region, R, in the plane was rotated about the x-axis to obtain a solid whose volume is given by

a. $\int_0^{\pi/2} \cos^2 x \, dx$.

b. $\int_{-1}^3 \pi (x^2 + 1) \, dx$.

In each case, what is the upper boundary of R and what are the limits?

Solution: In general, the volume is given by $\int_a^b \pi (f(x))^2 \, dx$.

a.   Here, $\pi (f(x))^2 = \cos^2 x$, so that $f(x) = \cos x/\sqrt{\pi}$. Hence, R is the region between the curve $y = \cos x/\sqrt{\pi}$ and the x-axis, $0 \le x \le \pi/2$.

b.   Set $\pi (f(x))^2 = \pi (x^2 + 1)$. Solving for f, we obtain $f(x) = \sqrt{x^2 + 1}$. R is then the region below this curve, for $-1 \le x \le 3$.

5.7   Let R be the region bounded by the curves $y = x$ and $y = x^2$. If R is rotated about the x-axis, find the volume of the solid generated.

Solution: A sketch of the region R is given in Figure 4-29. The two curves intersect at $x = 0$ and $x = 1$, which become the limits of integration. As in Solved Problem 4.12 (which dealt with the area between two curves), the volume sought here can be found by subtracting the volume generated by revolving $y = x^2$ (the smaller of the two functions) about the x-axis from the volume generated by $y = x$. We thus obtain

$$\int_0^1 \pi x^2 \, dx - \int_0^1 \pi x^4 \, dx \quad = \quad \pi \left[\frac{x^3}{3} - \frac{x^5}{5} \right]\Big|_0^1$$

$$= \quad \frac{2\pi}{15}.$$

AVERAGE VALUE

We know how to compute the average of two numbers, c_1 and c_2: We just add them and divide by 2. More generally, the average of n numbers, c_1, c_2, ..., c_n, is simply

$$\frac{c_1 + c_2 + \cdots + c_n}{n}.$$

But what can we possibly mean by the average value of a *function*? For example, suppose you read in the newspaper that the average temperature for October 15th was $57°$ F. What is meant by this number? I know the answer to this question, but I'll keep you in the dark while we develop a *calculus* approach.

To compute the average temperature, we could take readings every hour, and average those 24 numbers. If we did this, we would say that the average temperature, T_{avg} is given by

$$T_{avg} = \frac{T_1 + T_2 + \cdots + T_{24}}{24}, \tag{5.13}$$

where T_i is the temperature at hour i, $1 \le i \le 24$. (By 24 we mean midnight.) Or, if we wish to be more precise, we can take readings every half hour (48 in all), and then we would have

$$T_{avg} = \frac{T_{1/2} + T_1 + T_{3/2} + T_2 + \cdots + T_{47/2} + T_{24}}{48}. \tag{5.14}$$

(Here, $T_{1/2}$ is the temperature at hour $1/2$ (12:30 A.M.), T_1 the temperature at 1:00 A.M., and so forth.) But T_{avg} computed from (5.14) will probably be different from that given by (5.13), so what we really have in (5.13) and (5.14) are *approximations* to the average temperature. In line with our usual approach,

$\mathcal{A}$

we now proceed to improve these approximations. But first, we must assume that we have a *complete* picture of the temperature at *any* time of the day or night. In other words, we have a *function $T(t)$* which gives the temperature at any time t, $0 \le t \le 24$. So (5.13) and (5.14) can be rewritten as

$$\sum_{i=1}^{24} T(i)\frac{1}{24} \quad \text{and} \quad \sum_{i=1}^{48} T\left(\frac{i}{2}\right)\frac{1}{48},$$

respectively. If we take readings every quarter hour, we obtain

$$\sum_{i=1}^{96} T\left(\frac{i}{4}\right)\frac{1}{96}$$

as another approximation to T_{avg}.

Let's now become completely general. Divide the interval $[0, 24]$ into n equal subintervals and sample the temperature once on each subinterval at an arbitrary time, c_i, in the ith subinterval. We obtain

$$\sum_{i=1}^{n} T(c_i)\frac{1}{n}$$

as an approximation to T_{avg}. By letting n become larger, we *refine* the approximation. However (and here comes the clever step!),

$$\begin{aligned}
\sum_{i=1}^{n} T(c_i)\frac{1}{n} &= \sum_{i=1}^{n} \frac{T(c_i)}{n}\frac{24}{24} \\
&= \frac{1}{24}\sum_{i=1}^{n} T(c_i)\frac{24}{n} \\
&= \frac{1}{24}\sum_{i=1}^{n} T(c_i)\Delta t_i,
\end{aligned}$$

since $24/n = \Delta t_i$ is the length of each subinterval.

So we find that T_{avg} is approximately equal to

$$\frac{1}{24}\sum_{i=1}^{n} T(c_i)\Delta t_i,$$

which (except for the factor $1/24$) is a *Riemann sum* for the function $T(t)$ on $[0, 24]$. To obtain the exact value of T_{avg}, we pass to the *limit*. Thus,

$$T_{avg} = \lim_{n \to \infty} \frac{1}{24}\sum_{i=1}^{n} T(c_i)\Delta i = \frac{1}{24}\int_{0}^{24} T(t)\, dt.$$

Now, is this how the weather bureau calculates the average temperature? No. They just average the highest and lowest temperatures of the day!

In general, we define the *average value* of a function f on $[a, b]$ to be

$$f_{avg} = \frac{1}{b-a} \int_a^b f(x)\, dx.$$

5.8   Find the average value of $f(x) = x^2$ on $[2, 5]$.

Solution: Since the length of the interval of integration is 3,

$$f_{avg} = \frac{1}{3} \int_2^5 x^2\, dx = \left. \frac{x^3}{9} \right|_2^5 = \frac{1}{9}(125 - 8) = 13.$$

(Note that if we use the weather bureau approach — the average of the highest and lowest values — we'd get $(4 + 25)/2 = 14.5$.)

5.9   Find the average value of $f(x) = \sin^2 x$ on $[0, 2\pi]$.

Solution:

$$
\begin{aligned}
f_{avg} &= \frac{1}{2\pi} \int_0^{2\pi} \sin^2 x\, dx \\
&= \frac{1}{2\pi} \int_0^{2\pi} \frac{1 - \cos 2x}{2}\, dx \\
&= \left. \frac{1}{4\pi} \left(x - \frac{\sin 2x}{2} \right) \right|_0^{2\pi} \\
&= \frac{1}{2}.
\end{aligned}
$$

5.10   Suppose that $f(x) \geq 0$ on $[a, b]$ and is concave up there. Show that

$$f_{avg} \leq \frac{f(a) + f(b)}{2}.$$

(In other words, the weather bureau 'average' is greater than our definition of the average value.)

Solution: Draw the line, L, connecting the points $(a, f(a))$ and $(b, f(b))$ (Figure 5-15). It is clear from Figure 5-15 that the area under L is greater than that under $y = f(x)$, so let's compute these two areas. The area under $y = f(x)$, of course, is equal to $\int_a^b f(x)\, dx$. The area under L is a trapezoid of height $(b - a)$ and bases $f(a)$ and $f(b)$. (Note that the trapezoid is 'standing on its

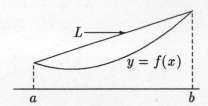

Figure 5-15: Concave up function

side' rather than in its usual orientation with the parallel lines horizontal.) The general formula for the area of a trapezoid of height h and bases b_1 and b_2 is

$$h\left(\frac{b_1 + b_2}{2}\right),$$

which in our case yields

$$(b-a)\left(\frac{f(a) + f(b)}{2}\right).$$

Thus,

$$\int_a^b f(x)\,dx \quad \leq \quad (b-a)\left(\frac{f(a) + f(b)}{2}\right), \quad \text{or}$$

$$\frac{1}{(b-a)}\int_a^b f(x)\,dx \quad \leq \quad \frac{f(a) + f(b)}{2},$$

which is equivalent to

$$f_{avg} \leq \frac{f(a) + f(b)}{2}.$$

MISCELLANEOUS APPLICATIONS

5.11   If the region in Figure 5-16 is revolved about the x-axis, a solid is formed. An approximation to the *surface area* (excluding the ends) of this solid is given by

$$\sum_{i=1}^n 2\pi f(c_i)\sqrt{1 + (f'(c_i))^2}\,\Delta x_i. \tag{5.15}$$

a.   Write down an integral which is equal to the *exact* value of the surface area.

b.   Use this result to find the surface area of the solid generated when the curve $y = x^3$, $0 \leq x \leq 1$, is revolved about the x-axis.

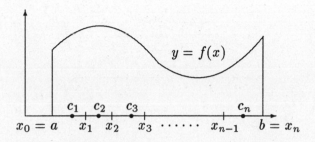

Figure 5-16: Surface area of a solid

Solution:

a.   The integral we want is obtained from (5.15), which is a Riemann sum. Its limit (as the mesh of the partition approaches 0) is

$$\int_a^b 2\pi f(x)\sqrt{1 + (f'(x))^2}\, dx.$$

b.   With $f(x) = x^3$, we obtain

$$\int_a^b 2\pi f(x)\sqrt{1 + (f'(x))^2}\, dx = \int_0^1 2\pi x^3 \sqrt{1 + 9x^4}\, dx.$$

Now

$$\frac{\pi}{27}(1 + 9x^4)^{3/2}$$

is an antiderivative of $2\pi x^3 \sqrt{1 + 9x^4}$. (Check this by differentiating!) Hence,

$$\int_0^1 2\pi x^3 \sqrt{1 + 9x^4}\, dx \quad = \quad \left. \frac{\pi}{27}(1 + 9x^4)^{3/2} \right|_0^1$$

$$= \quad \frac{\pi}{27}\left(10^{3/2} - 1\right)$$

5.12   The surface area of a sphere of radius R is $4\pi R^2$. Derive this formula by integration.

Solution: Watch how the formula emerges almost by magic. The surface area of the sphere is given by

$$\int_a^b 2\pi f(x)\sqrt{1 + (f'(x))^2}\, dx,$$

where

$$f(x) = \sqrt{R^2 - x^2}, \quad f'(x) = \frac{-x}{\sqrt{R^2 - x^2}}.$$

Thus, the area is equal to

$$2\pi \int_{-R}^{R} \sqrt{R^2 - x^2} \sqrt{1 + \frac{x^2}{R^2 - x^2}} \, dx \quad = \quad 2\pi \int_{-R}^{R} \sqrt{R^2 - x^2} \sqrt{\frac{R^2}{R^2 - x^2}} \, dx$$

$$= \quad 2\pi \int_{-R}^{R} R \, dx$$

$$= \quad 4\pi R^2.$$

5.13   Let R be the region bounded by the curves $y = 2x$ and $y = 4x - x^2$.

    a.   Sketch the region R.

    b.   Find the area of R.

    c.   Find the volume of the solid generated when R is revolved about the x-axis.

    d.   Write down (but do not evaluate) an expression involving integrals for the entire arc length of the boundary of R.

Solution:

    a.   The sketch appears in Figure 5-17.

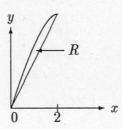

Figure 5-17: Region between two curves

    b.   The two curves intersect at $(0,0)$ and $(2,4)$, and $y = 4x - x^2$ lies above $y = 2x$ in this interval. Hence, the area of R is equal to

$$\int_{0}^{2} (4x - x^2 - 2x) \, dx \quad = \quad \int_{0}^{2} (2x - x^2) \, dx$$

$$= \quad \left(x^2 - \frac{x^3}{3} \right) \Bigg|_{0}^{2}$$

$$= \quad \frac{4}{3}.$$

c. The volume is equal to the *difference* of the volumes of two solids, the first generated by revolving $y = 4x - x^2$ about the x-axis, and the second obtained by revolving $y = 2x$. The resulting volume thus equals

$$\pi \int_0^2 \left[(4x - x^2)^2 - 4x^2 \right] dx \quad = \quad \pi \int_0^2 (12x^2 - 8x^3 + x^4) \, dx$$

$$= \quad \left(4x^3 - 2x^4 + \frac{x^5}{5} \right)\Big|_0^2$$

$$= \quad 6.4.$$

d. For the upper boundary, $f(x) = 4x - x^2$, so that $f'(x) = 4 - 2x$ and $(f'(x))^2 = 4x^2 - 16x + 16$. Hence, the length of the upper boundary is

$$\int_0^2 \sqrt{1 + (f'(x))^2} \, dx = \int_0^2 \sqrt{4x^2 - 16x + 17} \, dx.$$

The lower boundary is a straight line, whose length is $\sqrt{20} = 2\sqrt{5}$. The total length is thus

$$\int_0^2 \sqrt{4x^2 - 16x + 17} \, dx + 2\sqrt{5}.$$

5.14 Integrals have many *interpretations*. Consider, for example, $\int_0^1 \sqrt{x^2 + 1} \, dx$.

a. Find a region of the plane whose *area* is equal to this integral.

b. Find a region of the plane which, when revolved about the x-axis, generates a solid whose *volume* is equal to this integral.

c. Find a curve whose *length* is equal to this integral.

Solution: In each of the first two parts, the left- and right-hand boundaries are $x = 0$ and $x = 1$, and the lower boundary is the x-axis.

a. The region is bounded above by the curve $y = \sqrt{x^2 + 1}$.

b. The volume is given by the general formula

$$\pi \int_0^1 (f(x))^2 \, dx.$$

We set

$$\pi (f(x))^2 = \sqrt{x^2 + 1} = (x^2 + 1)^{1/2}.$$

Thus,

$$(f(x))^2 = \frac{(x^2 + 1)^{1/2}}{\pi},$$

or

$$f(x) = \frac{(x^2+1)^{1/4}}{\sqrt{\pi}},$$

so that the upper boundary of the region is the curve

$$y = \frac{(x^2+1)^{1/4}}{\sqrt{\pi}}.$$

c.   Here we set

$$\sqrt{1 + (f'(x))^2} = \sqrt{x^2 + 1},$$

or

$$1 + (f'(x))^2 = x^2 + 1.$$

This yields $f'(x) = x$ so that $f(x) = x^2/2$, and the curve has the equation $y = x^2/2, \ 0 \le x \le 1$.

Supplementary Problems

5.15   Find the length of the curve

$$y = \frac{1}{3}(x^2 + 2)^{3/2}$$

from $x = 1$ to $x = 4$.

5.16   Find the length of the curve

$$y = \frac{x^3}{6} + \frac{1}{2x}$$

from $x = 1$ to $x = 2$.

5.17   Set up the integral (do not evaluate it) which represents the length of one arch of the sine function; that is, which computes the length of the curve $y = \sin x, \ 0 \le x \le \pi$.

5.18   Find the average value of the following functions on the indicated intervals:

a. $\sin x$ on $[0, \pi]$ and on $[0, 2\pi]$.

b. $x^2 - 2x + 4$ on $[1, 3]$.

c. e^x on $[0, 1]$.

5.19 Show that if f is concave down, then

$$f_{avg} \geq \frac{f(a) + f(b)}{2}.$$

(The proof is similar to Solved Problem 5.10.)

5.20 In each part find the volume of the solid obtained by rotating the region R about the x-axis.

 a. R : bounded by $y = e^x$, the x-axis, and the lines $x = 0$ and $x = 1$.

 b. R : bounded by $y = \tan x$, the x-axis, $x = 0$ and $x = \pi/4$.

 c. R : bounded by the curve $y = \sec x$, the x-axis, and the lines $x = -\pi/4$ and $x = \pi/4$.

 d. R is the region between $y = 2 - x^2$ and $y = 1$.

Answers to Supplementary Problems

5.15 24

5.16 17/12

5.17 $\displaystyle\int_0^{\pi} \sqrt{1 + \cos^2 x}\, dx$

5.18 **a.** On $[0, \pi]$: $2/\pi$; on $[0, 2\pi]$: 0 **b.** 13/3 **c.** $e - 1$

5.20 **a.** $\pi\dfrac{(e^2 - 1)}{2}$ **b.** $\pi\left(1 - \dfrac{\pi}{4}\right)$ **c.** 2π **d.** $\dfrac{56\pi}{15}$

Chapter 6

Topics in Integration

In this chapter we study two topics involving the integral. The first is an extension of the basic concept to what is known as an *improper integral*. The second is called *numerical integration*, or, "What to do if you can't find an antiderivative." In each case we extend well-understood ideas to new situations, which is another of the themes that repeats through much of calculus.

6.1   Improper Integrals

> **What We Know:** $\int_a^b f(x)\,dx$ for *bounded* functions on *bounded* intervals.
>
> **What We Want To Know:** $\int_a^b f(x)\,dx$ for an *unbounded* function, f, and $\int_a^\infty f(x)\,dx$; $\int_{-\infty}^b f(x)\,dx$; $\int_{-\infty}^\infty f(x)\,dx$ for a *bounded* function, f.
>
> **How We Do It:** Approximate the *improper integrals* with integrals of *bounded* functions on *bounded* intervals.

In our discussion of the integral, you may recall that we considered only *bounded* functions on *bounded* intervals. In other words, $\int_a^b f(x)\,dx$ is defined (up to this point) only for functions satisfying $m \leq f(x) \leq M$, for some real numbers, m and M, and only when a and b are real numbers. Thus, the symbols

$$\int_0^1 \frac{1}{\sqrt{x}}\,dx, \quad \int_0^1 \frac{1}{x^2}\,dx \text{ and } \int_1^\infty \frac{1}{x^3}\,dx,$$

for example, are not yet meaningful, the first two because the *functions* are unbounded (the denominators are zero at $x = 0$), the third because the *interval of integration* is infinite in length. Our purpose in this section is to provide that meaning where possible. We will see that the first and third of the above integrals are meaningful, in an extended sense, but not the second one.

Now, hold on a second. Doesn't this problem seem awfully artificial, the kind of game that mathematicians invent when they're bored? Really, why should we be interested in it? Well, the answer is that these types of integrals arise in many vital, real-life situations. For example, we mentioned in Chapter 1 that the Fast Fourier Transform (FFT) involves an important application of the integral. A Fourier transform, in fact,

150

is actually an *improper* integral, while the FFT is a method for quickly computing Fourier transforms. And yet, this concept is so important, with so many widespread applications, that by now it has become a standard part of the engineering curriculum, even at the undergraduate level. In particular, the FFT makes possible the analysis of large amounts of data in what is known as *real time*. For example, it can be used to detect an incoming missile *before* it lands, thereby allowing it to be shot down. Also, it rapidly supplies vital information to a cardiologist about the condition of a heart attack victim, information that can sometimes mean the difference between life and death. So a topic, seemingly *invented* by mathematicians, turns out to have far-flung consequences.

Let's look at this problem geometrically for a moment. Recalling the important connection between the integral and area, we ask: Can an *unbounded* region of the plane have finite area? Let's elaborate.

Consider the graph of the function of our first example, $f(x) = 1/\sqrt{x}$, $0 < x \le 1$ (Figure 6-1(a)). Since $\lim_{x \to 0^+} f(x) = \infty$, the shaded region, R, which lies under the

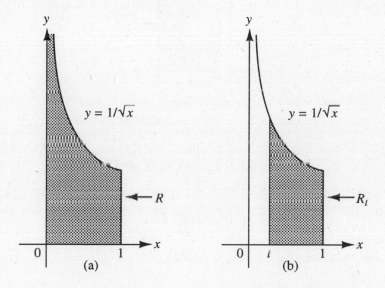

Figure 6-1: The region and an initial approximation

curve $y = 1/\sqrt{x}$, $0 < x \le 1$, is *unbounded*. (**Note:** The symbol $\lim_{x \to 0^+} f(x)$ means that x is approaching 0 from the *right*. Similarly, $\lim_{x \to 0^-} f(x)$ means the limit as x approaches 0 from the *left*.) Nevertheless, is it possible for the area of R to be *finite*? At first glance, the answer seems obviously to be no, since the region extends indefinitely in the y-direction. But let's proceed (perhaps naively) for a moment and consider the following approach:

Assuming that the region *does* have finite area, how can we find it? Let's apply our usual method of Approximation — Refinement — Limit ($\mathcal{A} - \mathcal{R} - \mathcal{L}$). A natural way to *approximate* $\int_0^1 f(x)\,dx$ is to consider not the full interval $[0,1]$, but rather a

$\mathcal{A}$

$\mathcal{R}$

subinterval $[t, 1]$, where t is 'close to' 0.   Denote by R_t the region under the curve $y = 1\sqrt{x}$, $t \leq x \leq 1$ (Figure 6-1(b)). We think of the area of R_t as an *approximation* to the area of R. However, since $f(x) = 1/\sqrt{x}$ is a *bounded function* on the subinterval $[t, 1]$, this area can be expressed as the ordinary integral, $\int_t^1 (1/\sqrt{x}) \, dx$.   Now, how can we improve or *refine* this approximation? Clearly, by moving t closer to 0 (Figure 6-2). Finally, to find the exact area of R, we pass to the *limit*:

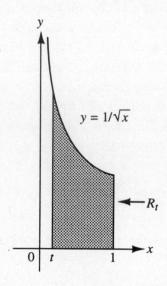

Figure 6-2: A refined approximation

$\mathcal{L}$

$$\text{area of } R = \lim_{t \to 0^+} \text{area of } R_t,$$

or

$$\text{area of } R = \lim_{t \to 0^+} \int_t^1 \frac{1}{\sqrt{x}} \, dx.$$

It is natural to use the integral sign as a notation for the area of R. Hence, we define

$$\int_0^1 \frac{1}{\sqrt{x}} \, dx = \lim_{t \to 0^+} \int_t^1 \frac{1}{\sqrt{x}} \, dx.$$

To compute the area of R we evaluate $\int_t^1 (1/\sqrt{x}) \, dx$ (most easily by using the Fundamental Theorem of Calculus, page 111), which gives us a function of t. We then take the limit of this function as $t \to 0^+$. Specifically,

$$\int_t^1 \frac{1}{\sqrt{x}} \, dx = 2\sqrt{x} \Big|_t^1 = 2 - 2\sqrt{t},$$

so that

$$\int_0^1 \frac{1}{\sqrt{x}} \, dx = \lim_{t \to 0^+} \left(2 - 2\sqrt{t}\right) = 2.$$

We now define the *improper integral*. Suppose that $\lim_{x \to a+} f(x) = \infty$, but that f is bounded on any subinterval of $(a, b]$ which does not contain a. If

$$\lim_{t \to a+} \int_t^b f(x)\, dx \tag{6.1}$$

exists, then we say that $\int_a^b f(x)\, dx$ *converges* and is equal to this limit. If the limit fails to exist, then $\int_a^b f(x)\, dx$ *diverges*. (A similar definition exists if the 'problem point' is b, with the limit then taken from the *left*.)

We turn to the next of our examples, $f(x) = 1/x^2$ on $(0,1]$. By our definition, the improper integral of $1/x^2$ converges on $(0,1]$ *if*

$$\lim_{t \to 0+} \int_t^1 \frac{1}{x^2}\, dx$$

exists. *However*, by the Fundamental Theorem of Calculus,

$$\int_t^1 \frac{1}{x^2}\, dx = \left. \frac{-1}{x} \right|_t^1 = \frac{1}{t} - 1,$$

so that

$$\lim_{t \to 0+} \int_t^1 \frac{1}{x^2}\, dx = \lim_{t \to 0+} \left(\frac{1}{t} - 1 \right),$$

which *does not exist*, since $1/t$ tends to ∞ as $t \to 0^+$.

There is a second type of improper integral, in which the function is bounded, but the *interval of integration* is unbounded (Figure 6-3). We now wish to make sense of the

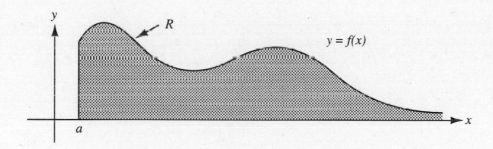

Figure 6-3: An unbounded region

expression $\int_a^\infty f(x)\, dx$. As usual, approximations will play an important role. Suppose, initially, that $f(x) \geq 0$ on $[a, \infty)$, so that we can use area considerations again. This time we cut off the region at some point, t, where t is large, and find the area of the subregion, R_t, under the curve from a to t (Figure 6-4). But this area is simply equal to $\int_a^t f(x)\, dx$, and here there are no problems, since both the function and interval are bounded. We next *refine* the approximation by moving t to the right (Figure 6-5). Finally, we obtain the precise value of the full area by taking the *limit*. Hence,

$$\int_a^\infty f(x)\, dx = \lim_{t \to \infty} \int_a^t f(x)\, dx, \tag{6.2}$$

provided this limit exists.

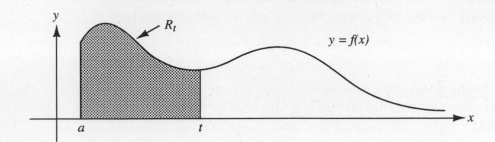

Figure 6-4: An initial approximation

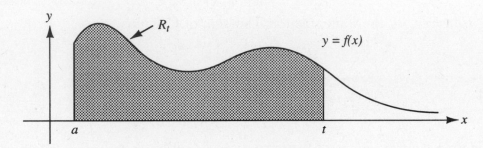

Figure 6-5: A better approximation

Example 6.1

$$
\begin{aligned}
\int_1^\infty \frac{1}{x^3}\,dx &= \lim_{t\to\infty} \int_1^t \frac{1}{x^3}\,dx \\[2mm]
&= \lim_{t\to\infty} -\frac{1}{2x^2}\bigg|_1^t \\[2mm]
&= \lim_{t\to\infty} \left(\frac{1}{2} - \frac{1}{2t^2}\right) \\[2mm]
&= \frac{1}{2},
\end{aligned}
$$

while

$$
\begin{aligned}
\int_1^\infty (1/\sqrt{x})\,dx &= \lim_{t\to\infty} \int_1^t (1/\sqrt{x})\,dx \\[2mm]
&= \lim_{t\to\infty} 2\sqrt{x}\,\bigg|_1^t \\[2mm]
&= \lim_{t\to\infty} 2\sqrt{t} - 1,
\end{aligned}
$$

which does not exist. Hence, $\int_1^\infty 1/(x^3)\,dx$ exists, but $\int_1^\infty (1/\sqrt{x})\,dx$ does not.

Just as the interval of integration can extend indefinitely to the right, it can do so to the left. We define such an integral in the obvious way:

$$\int_{-\infty}^{b} f(x)\,dx = \lim_{t \to -\infty} \int_{t}^{b} f(x)\,dx. \tag{6.3}$$

Example 6.2 *Consider* $f(x) = e^x$ *on* $(-\infty, 0]$ (Figure 6-6).

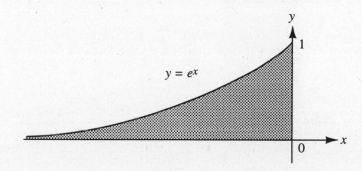

Figure 6-6: The graph of $y = e^x$ on $(-\infty, 0]$

Solution:

$$
\begin{aligned}
\int_{-\infty}^{0} e^x\,dx &= \lim_{t \to -\infty} \int_{t}^{0} e^x\,dx \\
&= \lim_{t \to -\infty} e^x \Big|_{t}^{0} \\
&= \lim_{t \to -\infty} (e^0 - e^t) \\
&= 1 - \lim_{t \to -\infty} e^t \\
&= 1.
\end{aligned}
$$

What about 'two-sided' infinite integrals? A possible way of defining such an integral is symmetrically, integrating from $-t$ to t and letting $t \to \infty$:

$$\int_{-\infty}^{\infty} f(x)\,dx = \lim_{t \to \infty} \int_{-t}^{t} f(x)\,dx. \tag{6.4}$$

However, this definition is generally rejected (although it has value and is used in certain advanced mathematics courses) because we consider each half of the interval to be a 'problem,' and we require that *each* such problem be resolved separately. So we split the integral into two integrals by choosing *any* intermediate point (say 0) and writing

$$\int_{-\infty}^{\infty} f(x)\,dx = \int_{-\infty}^{0} f(x)\,dx + \int_{0}^{\infty} f(x)\,dx, \tag{6.5}$$

and say that $\int_{-\infty}^{\infty} f(x)\, dx$ *exists* (or converges) if and only if *both* of the integrals on the right side of (6.5) exist. If at least one of the one-sided infinite integrals fails to exist, then we say that $\int_{-\infty}^{\infty} f(x)\, dx$ does not exist either.

Example 6.3　*Does $\int_{-\infty}^{\infty} x\, dx$ converge?*

Solution: If we had chosen the originally proposed definition (6.4) of this type of improper integral,

$$\int_{-\infty}^{\infty} f(x)\, dx = \lim_{t \to \infty} \int_{-t}^{t} f(x)\, dx,$$

we would find that

$$
\begin{aligned}
\int_{-\infty}^{\infty} x\, dx &= \lim_{t \to \infty} \int_{-t}^{t} x\, dx \\[2mm]
&= \lim_{t \to \infty} \left. \frac{x^2}{2} \right|_{-t}^{t} \\[2mm]
&= \lim_{t \to \infty} \left(\frac{t^2}{2} - \frac{(-t)^2}{2} \right) \\[2mm]
&= 0
\end{aligned}
$$

(Figure 6-7). What is happening is that because of the symmetry of $f(x) = x$, the huge

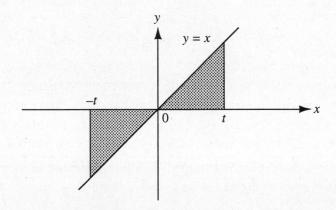

Figure 6-7: The graph of $y = x$ on $(-\infty, \infty)$

'positive area' between 0 and t is *exactly* cancelled by the huge 'negative area' between $-t$ and 0. According to definition (6.5) that we have adopted, however,

$$\int_{-\infty}^{\infty} x\, dx = \int_{-\infty}^{0} x\, dx + \int_{0}^{\infty} x\, dx,$$

and *neither* of the integrals on the right side exists. Hence, $\int_{-\infty}^{\infty} x\, dx$ diverges.

In a similar fashion, we separate any combination of 'problems,' whether there are two places where the function is unbounded, such as

$$\int_0^1 \frac{1}{x(1-x)}\, dx,$$

whose denominator is 0 at both 0 and 1, or one such place together with an infinite interval.

Example 6.4 *Consider $\int_0^\infty (1/\sqrt{x})\, dx$.*

Solution: There are two problems here: The integrand is unbounded in the vicinity of 0, and the interval of integration is infinite (Figure 6-8). So we break up the interval in

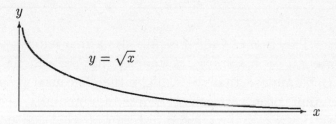

$$y = \sqrt{x}$$

Figure 6-8: A function with two problems

order to separate the problems:

$$\int_0^\infty (1/\sqrt{x})\, dx = \int_0^1 (1/\sqrt{x})dx + \int_1^\infty (1/\sqrt{x})\, dx.$$

In the first of the integrals on the right, the interval is bounded, but the function is unbounded, so that we have our first type of improper integral. In the second integral the situation is reversed: The function is bounded, but the interval of integration is unbounded. But notice that each of these integrals is problematic in only one way. Now, $\int_0^1 (1/\sqrt{x})\, dx = 2$, as we saw at the very beginning of this chapter (Page 151). However, we showed in Example 6.1 on page 154 that $\int_1^\infty (1/\sqrt{x})\, dx$ does not exist. Hence, $\int_0^\infty (1/\sqrt{x})\, dx$ *does not exist* either.

There aren't many results of a general nature in this topic, but there is one class of functions which can be fully analyzed, namely, integrands of the form $f(x) = 1/(x^p)$.

Theorem:

(a) $\int_0^1 (1/x^p)\, dx$ exists if and only if $p < 1$.

(b) $\int_1^\infty (1/x^p)\, dx$ exists if and only if $p > 1$.

The proof is found in Solved Problems 6.1 and 6.2. Notice that a consequence of this theorem is that no function of the form $1/x^p$ is integrable on $(0, \infty)$, for any value of p.

6.2   Numerical Integration

What We Know:
(a) How to compute the definite integral of *linear* functions.
(b) How to compute the definite integral of *quadratic* functions.

What We Want To Know: How to compute the definite integral of *arbitrary* functions.

How We Do It:
(a) Approximate f with *piecewise linear* functions (*Trapezoidal Rule*).
(b) Approximate f with *quadratic* functions (*Simpson's Rule*).

The Fundamental Theorem of Calculus provides a marvelously simple way of evaluating a definite integral — assuming, of course, that we can find an antiderivative of the integrand. But suppose that we can't, either because we haven't yet learned the appropriate technique of integration or, more fundamentally, because the integrand has no antiderivative that can be expressed in terms of what are called the *elementary functions* (polynomials, rational, algebraic, trigonometric, exponential, and logarithmic functions). Examples of this type are numerous. Two simple cases are $\int \sin(x^2)\,dx$ and $\int e^{x^2}\,dx$, neither of which can be expressed as a combination of elementary functions. And yet, we may need to evaluate definite integrals involving functions of this sort. For example, $\int_a^b e^{-x^2}\,dx$ (or simple variants thereof) plays a vital role in probability and statistics, and precise values of this integral for various choices of a and b must be obtained. But how, since a closed-form antiderivative is not available? We need *numerical* techniques for *approximating* definite integrals, and this section is devoted to that task. It should be pointed out that in this section the approximations, themselves, are the *ultimate goal*, rather than just the first stage in a process, as in most of our previous work.

Since the subject of numerical integration is vast, we will confine our attention to two of the most basic (but highly useful) methods of approximation, the Trapezoidal Rule and Simpson's Rule.

6.2.1   Trapezoidal Rule

Suppose I told you that there is an efficient method for approximating the definite integral of *any* function which requires us to know nothing more than how to integrate *linear functions*. Would you believe it? Well, there is! And since it is trivial to integrate linear functions, we have a straightforward method for approximating integrals. Now for the details.

Suppose we wish to evaluate $\int_a^b f(x)\,dx$. We begin by partitioning $[a,b]$ by a number of *equally-spaced* points, $a = x_0, x_1, x_2, \ldots, x_n = b$ (Figure 6-9). Since there are

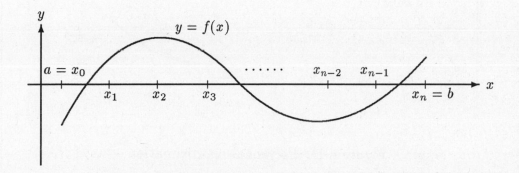

Figure 6-9: Partition of the interval

n points, the length of each subinterval is $(b-a)/n$, which we denote by h. Hence, $x_1 = a + h$, $x_2 = a + 2h$, $x_3 = a + 3h, \ldots, x_n = a + nh$. We now approximate the function f on the first subinterval, $[x_0, x_1]$, by passing a straight line through the points (x_0, y_0) and (x_1, y_1) on the graph of $y = f(x)$ (here $y_0 = f(x_0)$ and $y_1 = f(x_1)$) (Figure 6-10). We continue with similar approximations on each of the subintervals,

$\mathcal{A}$

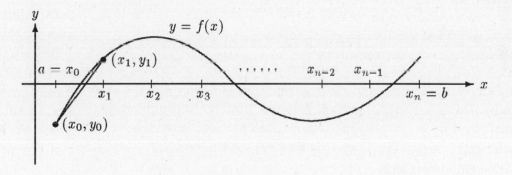

Figure 6-10: Approximation on first subinterval

$[x_1, x_2], [x_2, x_3], \ldots, [x_{n-1}, x_n]$, thereby obtaining a broken-line (polygonal) approximation, L, to f over the entire interval $[a,b]$ (Figure 6-11). Now, instead of computing $\int_a^b f(x)\,dx$, which we can't evaluate exactly because we can't find an antiderivative of $f(x)$, we compute $\int_a^b L(x)\,dx$. This is most easily accomplished by breaking down this integral into n separate integrals, one over each of the subintervals, $[x_0, x_1], [x_1, x_2], \ldots, [x_{n-1}, x_n]$. Let's see what we obtain. Assume, initially, that L is positive in a particular interval, say $[x_{i-1}, x_i]$ (Figure 6-12). Then $\int_{x_{i-1}}^{x_i} L(x)\,dx$ is just the area of a trapezoid (standing on its side, rather than in its usual orientation with the parallel sides horizontal), whose bases have lengths y_{i-1} and y_i, and whose height

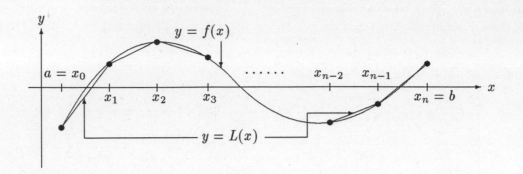

Figure 6-11: Polygonal approximation

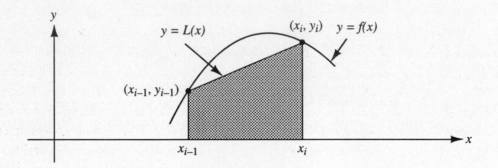

Figure 6-12: Area of a trapezoid

is $x_{i-1} - x_i = h$. The area of such a trapezoid is $h(y_{i-1} + y_i)/2$. Moreover, $\int_{x_{i-1}}^{x_i} L(x)\, dx$ is equal to $h(y_{i-1} + y_i)/2$, even if L is not positive throughout $[x_{i-1}, x_i]$. (This can be demonstrated by a simple, but tedious, calculation, in which we derive the *equation* of L on the interval $[x_{i-1}, x_i]$, and then integrate the result.)

We've thus seen that

$$\int_{x_{i-1}}^{x_i} L(x)\, dx = \frac{h(y_{i-1} + y_i)}{2}.$$

Since, as we've noticed, this result is valid on *any* interval, we can successively substitute $i = 1, 2, \ldots, n$, obtaining

$$\int_{x_0}^{x_1} L(x)\, dx = \frac{h(y_0 + y_1)}{2},$$

$$\int_{x_1}^{x_2} L(x)\, dx = \frac{h(y_1 + y_2)}{2},$$

$$\int_{x_2}^{x_3} L(x)\, dx = \frac{h(y_2 + y_3)}{2},$$

and so forth, until

$$\int_{x_{n-1}}^{x_n} L(x)\,dx = \frac{h(y_{n-1}+y_n)}{2}.$$

Adding all of these together we obtain

$$\int_a^b L(x)\,dx = \int_{x_0}^{x_1} L(x)\,dx + \int_{x_1}^{x_2} L(x)\,dx + \cdots + \int_{x_{n-1}}^{x_n} L(x)\,dx$$

$$= \frac{h(y_0+y_1)}{2} + \frac{h(y_1+y_2)}{2} + \cdots + \frac{h(y_{n-1}+y_n)}{2}$$

$$= \frac{h(y_0+2y_1+2y_2+\cdots+2y_{n-1}+y_n)}{2},$$

which we denote by T_n (T for trapezoid, n for the number of subintervals). Hence,

$$T_n = \frac{h(y_0+2y_1+2y_2+\cdots+2y_{n-1}+y_n)}{2} \tag{6.6}$$

is an *approximation* to $\int_a^b f(x)\,dx$. This approximation can be *refined* in the obvious way, by taking more subintervals and thereby making the spacing, h, smaller. In theory, we can obtain the exact value by passing to the *limit* (letting $n \to \infty$), but since, as we said earlier (Page 158), we are only seeking an estimate of the integral, we do not pursue this step. However, some natural questions remain: How good an estimate do we need? At what point should we stop refining the approximation and be satisfied with the current computation? The answers to these questions depend upon our ability to develop accurate error estimates or tolerances for T_n. We won't go through the details of such a calculation, which are found in most texts. Instead, we focus on what we should expect the error to depend on. It seems pretty clear that the spacing, h, will play an important role. As h gets smaller, then so does the error, since L is then a better approximation to f. However, we still have to determine the precise connection between h and the error. But h is not the only important feature to be analyzed. It seems possible that the trapezoidal method will work better (that is, be more efficient) for certain functions than for others. So we ask the following question, the answer to which will give us a vital clue in the general case: For which functions (if any) is the Trapezoidal Rule *exact* (that is, no error at all, regardless of the size of h)? A glance at Figure 6-11 on page 160 will convince us that this phenomenon will occur if f is a *linear function*, since the approximating function, L, obviously coincides with f in this case. Now, if the error is 0 for linear functions, then we might expect it to be *small* if f is 'nearly linear,' a concept we encountered previously in Chapter 3, page 64, where we analyzed the behavior of the error in linear approximation. We saw there that a 'nearly linear' function is one whose second derivative is 'small' everywhere. As a result of this discussion we can expect f'' to play a prominent part in the error estimate for T_n, and we now turn to the exact connection.

Let

$$E(T_n) = T_n - \int_a^b f(x)\,dx$$

be the error in the Trapezoidal Rule using n subintervals, and suppose that $|f''(x)| \leq M_2$ for all x in the interval $[a, b]$. Then

$$|E(T_n)| \leq \frac{M_2(b-a)^3}{12n^2} \tag{6.7}$$

Example 6.5 *Apply the Trapezoidal Rule to the integral*

$$\int_0^1 e^{x^2}\, dx.$$

Solution: Here

$$
\begin{aligned}
f(x) &= e^{x^2} \\
f'(x) &= 2xe^{x^2} \\
f''(x) &= 4x^2 e^{x^2} + 2e^{x^2} = e^{x^2}(4x^2 + 2).
\end{aligned}
$$

On the interval $[0, 1]$,

$$e^{x^2} \leq e^1 = e,$$

and $4x^2 + 2 \leq 6$. Hence, $|f''(x)| \leq 6e$. From (6.7) we obtain

$$E(T_n) \leq \frac{6e}{12n^2} < \frac{1.36}{n^2}.$$

Thus, for example, if we use $n = 10$ subintervals, then the error $E(T_{10})$ is guaranteed to be no more than $1.36/100 = .0136$, while if we use $n = 50$ intervals, then $E(T_{50}) \leq 1.36/2500 < .0006$.

Remark 6.1 We said earlier that the size of h is important in determining a bound on $E(T_n)$, and yet h does not appear in the error estimate (6.7). It can, however, be introduced by slightly modifying the form of the error. Recall that $h = (b-a)/n$, so that $h^2 = (b-a)^2/n^2$. We can thus rewrite (6.7) in the following equivalent form which emphasizes the dependence of $E(T_n)$ on the size of the spacing, h, rather than on the number of intervals, n:

$$|E(T_n)| \leq \frac{M_2(b-a)h^2}{12}. \tag{6.8}$$

Example 6.6 *The natural logarithm function is usually defined as an integral:*

$$\ln x = \int_1^x \frac{1}{t}\, dt. \tag{6.9}$$

Although the actual calculation of logarithms is generally carried out by means of infinite series, let's see what we can accomplish from the integral definition. For example, we'll use the Trapezoidal Rule to compute $\ln 2 = \int_1^2 (1/t)dt$.

Solution: Here $f(t) = 1/t$, so that two differentiations yield $f''(t) = 2/t^3$, whose maximum value on [1,2] is 2, occurring at $t = 1$. Hence, $M_2 = 2$, so that

$$|E(T_n)| \leq \frac{2(b-a)^3}{12n^2} = \frac{1}{6n^2}.$$

We obtain the following table for various values of n (the actual value of $\ln 2$ is .6931471806 to 10 decimal places):

n	T_n	Error Bound	Actual Error
10	.6937714031	.0016666667	.0006242225
20	.6933033815	.0004166667	.0001562009
50	.6931721794	.0000666667	.0000249988
100	.6931534306	.0000166667	.0000062500
200	.6931487435	.0000041667	.0000015629
500	.6931474310	.0000006667	.0000002504

Notice that the actual error is considerably less than the guaranteed error bound (in fact, less than 40% of the latter). Recall, however, that in this case we have an independent way of calculating $\ln 2$ and hence $\int_1^2 (1/t)\, dt$. Generally, we do not have this luxury, so that we won't know the actual error and will have to rely upon the error bound guaranteed by (6.7). Of course, we want to keep n as small as possible to minimize the amount of computation, so we choose n large enough to guarantee that the error is less than what we can tolerate. In other words, if we are doing a practical calculation and can allow an error of up to, say, .0001, but no more, then we choose n large enough that $|E(T_n)| \leq .0001$. In our case, since $|E(T_n)| \leq 1/(6n^2)$, this requires

$$\frac{1}{6n^2} \leq .0001,$$

which is equivalent to each of the following inequalities:

$$\frac{1}{6n^2} \leq \frac{1}{10000}$$

$$6n^2 \geq 10000$$

$$n^2 \geq 10000/6$$

$$n \geq 41.$$

Thus $n = 41$ suffices to guarantee that the error does not exceed .0001. (Our earlier calculations indicate that the error is actually smaller than this or, equivalently, that an even smaller value of n would probably be adequate. But, as mentioned before, in most cases we have no way of knowing the true value.)

6.2.2   Simpson's Rule

Although the Trapezoidal Rule appears to be an effective method for estimating a definite integral, more efficient techniques exist. We now present Simpson's method, perhaps the one most commonly used. Since the ideas underlying this technique are similar to those we encountered in the Trapezoidal Rule, we can quickly outline the procedure and error estimate. The key difference between the two methods is in the initial approximation; the Trapezoidal Rule is based upon linear functions, while Simpson's Rule employs quadratics (parabolas).

Once again, we begin by partitioning the interval $[a, b]$ by a number of equally-spaced points, $a = x_0, x_1, x_2, \ldots, x_n = b$, as in Figure 6-9, page 159. (This time, however, we require that n be an *even* number.) Now $\int_a^b f(x)\, dx$ is a sum of integrals,

$$
\begin{aligned}
\int_a^b f(x)\, dx &= \int_{x_0}^{x_n} f(x)\, dx \\
&= \int_{x_0}^{x_2} f(x)\, dx + \int_{x_2}^{x_4} f(x)\, dx + \cdots + \int_{x_{n-2}}^{x_n} f(x)\, dx
\end{aligned}
$$

Let's look at one of these integrals, $\int_{x_{i-2}}^{x_i} f(x)\, dx$, where we make use of the three points (x_{i-2}, y_{i-2}), (x_{i-1}, y_{i-1}), and (x_i, y_i) (Figure 6-13). Just as *two* points determine a

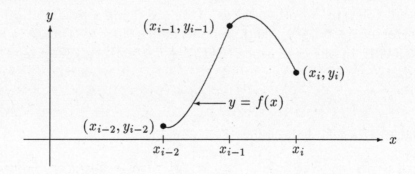

Figure 6-13: Three equally-spaced points

unique *line*, so any *three* points determine a unique *parabola* (or a line, if the three points happen to be collinear). We pass a parabola, p, (whose algebraic equation is a second degree polynomial, a quadratic) through the points (x_{i-2}, y_{i-2}), (x_{i-1}, y_{i-1}), and (x_i, y_i) (Figure 6-14). p serves as an approximation to f on $[x_{i-2}, x_i]$ and $\int_{x_{i-2}}^{x_i} p(x)\, dx$ as an *approximation* to $\int_{x_{i-2}}^{x_i} f(x)\, dx$. The computation of $\int_{x_{i-2}}^{x_i} p(x)\, dx$ is straight-forward, but somewhat involved, and, since it is found in your text, we just state the result here, namely,

$$
\int_{x_{i-2}}^{x_i} p(x)\, dx = \frac{h(y_{i-2} + 4y_{i-1} + y_i)}{3},
$$

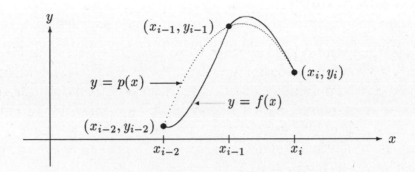

Figure 6-14: Simpson's approximation

where, as before, $h = (b - a)/n$, $y_{i-2} = f(x_{i-2})$, $y_{i-1} = f(x_{i-1})$, and $y_i = f(x_i)$. Because the index, i, is arbitrary here, we can apply this result to each of the intervals $[x_0, x_2], [x_2, x_4], \ldots, [x_{n-2}, x_n]$, (always taking 3 points at a time), and Simpson's Rule is then obtained by adding all of the approximations together. Specifically, we find

$$S_n = \frac{h(y_0 + 4y_1 + 2y_2 + 4y_3 + \cdots + 4y_{n-1} + y_n)}{3}. \tag{6.10}$$

Although the Simpson approximation is similar to the Trapezoidal Rule in form, it is actually far more powerful, as we'll see from the error estimates. Reasoning as we did earlier, we expect the error to depend upon h and the size of the third derivative of f, since the Simpson approximation is exact for all quadratic functions, $f(x) = a_2 x^2 + a_1 x + a_0$, whose third derivative is 0. In fact, we get an unexpected bonus: The approximation is also *exact* (error = 0) for cubic functions, $f(x) = a_3 x^3 + a_2 x^2 + a_1 x + a_0$, so that the error actually depends upon the *fourth* derivative of f. Specifically,

$$|E(S_n)| \leq \frac{M_4(b - a)^5}{180 n^4}, \tag{6.11}$$

where $M_4 = max|f''''(x)|$ for $a \leq x \leq b$. An alternative form for the error estimate, which emphasizes its dependence on the spacing h, is given by

$$|E(S_n)| \leq \frac{M_4(b - a)h^4}{180}. \tag{6.12}$$

Example 6.7 *Return to Example 6.6, to compute* $\ln 2 = \int_1^2 (1/t)\, dt$, *using Simpson's Rule.*

Solution: We obtain the table given below.

n	S_n	Error Bound	Actual Error
10	.6931502310	.0000133333	.0000030504
20	.6931473747	.0000008333	.0000001941
50	.6931471860	.0000000213	.0000000054

The errors in S_n are clearly much smaller than the corresponding ones in T_n. Additionally, if we want to guarantee an error of no more than, say, .0001, then the required value of n is much smaller for Simpson, as the following computation shows:

For $f(t) = 1/t$, we have $f''''(t) = 24/t^5$, so that $M_4 = \max |f''''(t)|$ for $1 \leq t \leq 2$ is 24. Hence,

$$|E(S_n)| \leq \frac{24(b-a)^5}{180n^4} = \frac{2}{15n^4}.$$

If we want $|E(S_n)| \leq .0001$, it suffices that

$$\frac{2}{15n^4} \leq .0001 = \frac{1}{10000}$$

or

$$\frac{15n^4}{2} \geq 10000$$

or

$$n^4 > \frac{20000}{15} = \frac{4000}{3}.$$

Thus, we can choose

$$n \geq \left(\frac{4000}{3}\right)^{1/4} = 6.043.$$

Since n must be even, $n = 8$ will suffice, while for the Trapezoidal Rule we saw earlier that $n = 41$ was required to guarantee this degree of accuracy.

Note: Do not conclude from the above discussion that Simpson's Rule is *always* more efficient than the Trapezoidal Rule. If f has a bounded fourth derivative, then Simpson's Rule will usually be more accurate than the Trapezoidal Rule, at least for n sufficiently large. This is so because the error using Simpson's Rule decreases like $1/n^4$, while that of the Trapezoidal Rule decreases like $1/n^2$. Even in this case, however, there are functions for which the Trapezoidal Rule gives better approximations. For example, the Trapezoidal Rule is known to be exceptionally efficient for *periodic* functions, such as $\sin x$ or $\cos x$. In addition, if f does not have bounded derivatives of the required order, then we may not achieve the estimates we obtained in this section. We'll see an example of this phenomenon in Solved Problem 6.15.

Solved Problems

IMPROPER INTEGRALS

6.1   Show that

$$\int_0^1 \frac{dx}{x^p}$$

converges if and only if $p < 1$.

Solution: Since we're going to use the Fundamental Theorem of Calculus to help solve the problem, we need an antiderivative of $1/x^p$. We have

$$\int \frac{dx}{x^p} = \begin{cases} x^{1-p}/(1-p) & \text{if } p \neq 1 \\ \ln x & \text{if } p = 1. \end{cases}$$

Suppose first that $p \neq 1$. By the definition of the improper integral,

$$\begin{aligned} \int_0^1 \frac{dx}{x^p} &= \lim_{t \to 0^+} \int_t^1 \frac{dx}{x^p} \\ &= \lim_{t \to 0^+} \left. \frac{x^{1-p}}{1-p} \right|_t^1 \\ &= \lim_{t \to 0^+} \left(\frac{1}{1-p} - \frac{t^{1-p}}{1-p} \right). \end{aligned}$$

The problematic term is $t^{1-p}/(1-p)$. If $p < 1$, then $1 - p > 0$, and

$$\lim_{t \to 0^+} t^{1-p} = 0.$$

So in this case the improper integral converges to $1/(1-p)$. On the other hand, if $p > 1$, then $1 - p < 0$, so that the term t^{1-p} appears in the *denominator* of the fraction, and this term is *unbounded* as $t \to 0^+$. Hence, the integral diverges.

The only case remaining is $p = 1$, where

$$\begin{aligned} \lim_{t \to 0^+} \int_t^1 \frac{dx}{x} &= \lim_{t \to 0^+} \left. \ln x \right|_t^1 \\ &= \lim_{t \to 0^+} (\ln 1 - \ln t) \\ &= \lim_{t \to 0^+} -\ln t. \end{aligned}$$

The latter term is unbounded as $t \to 0^+$, so the integral again diverges.

6.2   Show that

$$\int_1^\infty \frac{dx}{x^p}$$

converges if and only if $p > 1$.

Solution: The proof is similar to that of the previous problem. If $p \neq 1$, then

$$\int_1^\infty \frac{dx}{x^p} = \lim_{t\to\infty} \int_1^t \frac{dx}{x^p}$$

$$= \lim_{t\to\infty} \frac{x^{1-p}}{1-p}\bigg|_1^t$$

$$= \lim_{t\to\infty} \left(\frac{t^{1-p}}{1-p} - \frac{1}{1-p} \right).$$

This time, however, $p > 1$ is favorable, for then

$$\lim_{t\to\infty} \frac{t^{1-p}}{1-p} = 0,$$

and the integral is equal to $1/(p-1)$. If $p < 1$, then the term involving t^{1-p} tends to ∞ and the integral diverges. The case of $p = 1$ is left to the reader.

6.3   In some cases, it is possible to show that an improper integral converges, even though you can't find its exact value! In this problem we'll derive a result which allows us to compare two improper integrals and we'll see that the convergence or divergence of one of them can give us information about the other.

Prove the following result:

Comparison Theorem: Suppose $0 < f(x) \leq g(x)$ for all $x \geq a$.

a.   If $\int_a^\infty g(x)\,dx$ converges, then $\int_a^\infty f(x)\,dx$ also converges.

b.   If $\int_a^\infty f(x)\,dx$ diverges, then $\int_a^\infty g(x)\,dx$ also diverges.

Solution: The easiest way to understand this result is in terms of area (Figure 6-15).

a.   Suppose $\int_a^\infty g(x)\,dx$ converges. Since $g(x) > 0$, this means that the area of the region bounded above by $y = g(x)$, below by the x-axis and on the left by the line $x = a$ is *finite*. Now $f(x) \leq g(x)$, so the area of the comparable region below $y = f(x)$ is even smaller, and hence $\int_a^\infty f(x)\,dx$ also converges.

b.   Conversely, if $\int_a^\infty f(x)\,dx$ diverges, then the area of the region just mentioned is *not finite*, so that the area of the region below $y = g(x)$ is also not finite. Hence, $\int_a^\infty g(x)\,dx$ diverges.

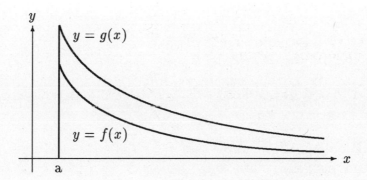

Figure 6-15: Comparison of two functions

6.4   Determine whether $\int_1^\infty e^{-x}/x\,dx$ converges or diverges.

Solution: e^{-x}/x is an example of a function which does not have an elementary antiderivative. In other words, there does not exist a combination of the basic functions of calculus whose derivative is equal to e^{-x}/x. As a result, we cannot use the Fundamental Theorem to solve this problem.

So what can we do? We fall back on the Comparison Theorem proved in 6.3. Of course, this theorem has a shortcoming: We aren't able to *evaluate* the integral. But we don't have a choice in the matter, and often the convergence or divergence of the integral is more important than the actual value.

To use the Comparison Theorem, we need a second integral to compare with the given one. In this case a good choice for y is e^{-x}. Since we're on the interval $[1, \infty)$, the integrand $e^{-x}/x \leq e^{-x}$. So if we can show that $\int_1^\infty e^{-x}\,dx$ converges, then the Comparison Theorem tells us that our integral also converges. Now,

$$
\begin{aligned}
\int_1^\infty e^{-x}\,dx &= \lim_{t\to\infty} \int_1^t e^{-x}\,dx \\
&= \lim_{t\to\infty} -e^{-x}\Big|_1^t \\
&= \lim_{t\to\infty}(-e^{-t} + e^{-1}) \\
&= e^{-1},
\end{aligned}
$$

since $e^{-t} = 1/e^t$ tends to 0 as t tends to ∞. Thus, $\int_1^\infty e^{-x}/x\,dx$ converges.

6.5   Investigate the convergence of $\int_1^\infty \sin^2 x/x^2\,dx$.

Solution: We use the Comparison Theorem again.

$$
0 \leq \frac{\sin^2 x}{x^2} \leq \frac{1}{x^2},
$$

and $\int_1^\infty 1/x^2 \, dx$ converges by Solved Problem 6.2.

6.6 Does $\int_1^\infty 1/\sqrt{x^3 + 1} \, dx$ converge?

Solution:

$$\frac{1}{\sqrt{x^3 + 1}} \leq \frac{1}{\sqrt{x^3}} = \frac{1}{x^{3/2}}.$$

By Solved Problem 6.2, we know that $\int_1^\infty 1/(x^{3/2}) \, dx$ converges. Thus, our integral converges as well.

6.7 Let

$$f(x) = \begin{cases} 1/\sqrt{x}, & 0 < x \leq 1 \\ 1/x^2, & 1 < x. \end{cases}$$

Evaluate $\int_0^\infty f(x) \, dx$.

Solution: The function has two problems: It is unbounded near 0, and the interval of integration is infinite. As we saw, in such a situation we must isolate the problems by breaking the interval into two pieces, as follows.

$$\int_0^\infty f(x) \, dx \;\; = \;\; \int_0^1 f(x) \, dx + \int_1^\infty f(x) \, dx$$

$$= \;\; \int_0^1 \frac{1}{\sqrt{x}} \, dx + \int_1^\infty \frac{1}{x^2} \, dx.$$

The value of the first of these integrals was computed at the very beginning of Chapter 6, and was found to be 2. The second integral can be evaluated in the usual way; it is equal to 1. Thus, our integral *exists*, and is equal to 3, the sum of the two integrals.

6.8 Does $\int_0^1 e^x/x \, dx$ converge?

Solution: $e^x/x \geq 1/x$ for $0 \leq x \leq 1$. We know from Solved Problem 6.1 that $\int_0^1 1/x \, dx$ diverges and the Comparison Test tells us that the present integral diverges as well.

6.9 A function, f, is said to be *odd* if $f(-x) = -f(x)$ for all x. (Examples of odd functions include all power functions, x^n, where n is an odd integer.) Suppose f is an odd function such that $\int_0^\infty f(x) \, dx$ converges. Prove that $\int_{-\infty}^\infty f(x) \, dx$ also converges and equals 0.

Solution: We saw that an integral from $-\infty$ to ∞ has to be broken into two separate integrals. A convenient breakpoint in this case is 0. Now we know that

$\int_0^\infty f(x)\,dx$ converges, say to a value L, and since f is odd

$$\int_{-\infty}^0 f(x)\,dx = \lim_{t\to-\infty}\int_t^0 f(x)\,dx$$

$$= \lim_{t\to\infty}\int_0^t -f(x)\,dx$$

$$= -L.$$

So the integral from $-\infty$ to 0 also exists, which tells us that $\int_{-\infty}^\infty f(x)\,dx$ exists and equals $L - L = 0$.

NUMERICAL INTEGRATION

6.10 Integrals of the form

$$\int_a^b e^{-x^2}\,dx$$

are of great importance in statistics. Unfortunately, however, they cannot be evaluated using the Fundamental Theorem, since the integrand does not have an elementary antiderivative. As a result, we have to rely upon a numerical approximation. Find an approximation to $\int_0^1 e^{-x^2}\,dx$, using Simpson's Rule with $n = 4$ subintervals.

Solution: Divide $[0, 1]$ into 4 equal subintervals. Hence, $x_0 = 0$, $x_1 = .25$, $x_2 = .5$, $x_3 = .75$ and $x_4 = 1$. Let $y_i = e^{-x_i^2}$, for $i = 0, 1, 2, 3, 4$. From (6.10), page 165,

$$S_4 = \frac{h(y_0 + 4y_1 + 2y_2 + 4y_3 + y_4)}{3},$$

where $h = .25$. Carrying out the evaluations of the function, we obtain $S_n = .25(1 + 3.757652 + 1.557602 + 2.279131 + .367879)/3 = .746855$.

6.11 How many subintervals are needed to guarantee an error not exceeding .001 when computing an approximation to $\int_1^3 \ln x\,dx$ using the Trapezoidal Rule?

Solution: In order to use the error estimate we must compute bounds on the size of the second derivative of $\ln x$. Letting $f(x) = \ln x$, we have

$$f'(x) = \frac{1}{x}, \quad f''(x) = -\frac{1}{x^2}.$$

On the interval $[1, 3]$, the maximum absolute value of f'' occurs when $x = 1$, yielding $M_2 = \max|f''(x)| = 1$. Now (6.7), page 162, tells us that the error in the Trapezoidal Rule when using n intervals satisfies

$$|E(T_n)| \leq \frac{M_2(b-a)^3}{12n^2} \leq \frac{8}{12n^2} = \frac{2}{3n^2},$$

since $M_2 = 1$ and $(b-a) = 2$. Solving the inequality

$$\frac{2}{3n^2} \leq .001$$

yields $n > 25.9$ or, since n must be an integer, 26 intervals suffice.

6.12   Repeat the previous problem for Simpson's Rule.

Solution: This time we need a bound on $|f''''|$. We have

$$f'''(x) = \frac{2}{x^3}, \quad f''''(x) = -\frac{6}{x^4}.$$

As in the previous problem, the largest value of $|f''''|$ in the interval $[1,3]$ occurs at $x = 1$, so that $M_4 = 6$. We use (6.11), page 165, which tells us that

$$|E(S_n)| \leq \frac{M_4(b-a)^5}{180n^4} \leq \frac{6 \cdot 32}{180n^4} = \frac{16}{15n^4}.$$

Now solve the inequality

$$\frac{16}{15n^4} \leq .001$$

to obtain $n > 5.8$, so that 6 intervals guarantee the desired accuracy.

6.13   Since the derivative of the function $\tan^{-1} x$ is $1/(1+x^2)$, we have

$$\int_0^1 \frac{dx}{1+x^2} = \tan^{-1} x \Big|_0^1 = \frac{\pi}{4}.$$

Hence,

$$\pi = 4 \int_0^1 \frac{dx}{1+x^2} = \int_0^1 \frac{4dx}{1+x^2},$$

and the integral can be used to find a numerical approximation to π. (No, 22/7 just won't do!) Use the Trapezoidal Rule and Simpson's Rule with 8 subintervals to approximate π.

Solution: Here $h = 1/8$, $x_i = i/8$, and $y_i = 4/(1+x_i^2)$, $i = 0, 1, 2, \ldots, 8$. From (6.6), we have

$$
\begin{aligned}
T_8 &= [y_0 + 2(y_1 + y_2 + y_3 + y_4 + y_5 + y_6 + y_7) + y_8]/16 \\
&= \frac{1}{16}\left[4 + 2\left(\frac{256}{65} + \frac{64}{17} + \frac{256}{73} + \frac{16}{5} + \frac{256}{89} + \frac{64}{25} + \frac{256}{113}\right) + 2\right] \\
&= 3.138988495.
\end{aligned}
$$

Simpson's Rule, which we obtain from (6.10), yields

$$S_n = [y_0 + 4y_1 + 2y_2 + 4y_3 + 2y_4 + 4y_5 + 2y_6 + 4y_7 + y_8]/24$$

$$= \frac{1}{24}\left[4 + 4\frac{256}{65} + 2\frac{64}{17} + 4\frac{256}{73} + 2\frac{16}{5} + 4\frac{256}{89} + 2\frac{64}{25} + 4\frac{256}{113} + 2\right]$$

$$= 3.141592503.$$

Since $\pi = 3.141592654$ (to 9 decimal places), we see that Simpson's Rule is accurate to 6 decimal places, while the Trapezoidal Rule is correct to only 2 places (when rounded). (If you're impressed by the accuracy of Simpson's method, keep in mind that different and *much* more powerful techniques have been used recently to compute π to over 2 *billion* decimal places!!)

6.14 Let f be a function which has four bounded derivatives, so that the error estimates for T_n and S_n (6.7) and (6.11) are valid. Suppose that to refine the approximation we cut the spacing, h, in half, which is equivalent to doubling the number of subintervals, n. What effect do we anticipate this will have on the errors in the approximation of $\int_a^b f(x)\,dx$?

Solution: From (6.7) and (6.11),

$$|E(T_n)| \le \frac{M_2(b-a)^3}{12n^2}.$$

Doubling n requires that we look at the bound for E_{2n}, which is

$$|E(T_{2n})| \le \frac{M_2(b-a)^3}{12\cdot 4n^2} = \frac{1}{4}\frac{M_2(b-a)^3}{12n^2},$$

which is 1/4 the error bound for T_n. So we expect $|E(T_{2n})|$ to be *about* 1/4 as large as $|E(T_n)|$.

Note that we can't be sure that $E(T_{2n})$ is always *exactly* one fourth $E(T_n)$. The reason for this is that both of these inequalities are upper bounds for the *possible* error. As we have seen, the actual errors are often somewhat smaller. Nevertheless, the computation we have just made does provide us with a reasonable idea of what to expect from the approximation. Note, also, that doubling n is tantamount to doubling the amount of work that we must do. The payoff, however, is that the estimate is about 4 times as accurate.

Satisfying as this result may be, applying similar analysis to Simpson's Rule gives an even more dramatic outcome. Here, doubling n leads to the error being decreased by a factor of 16! This is due to the presence of the term n^4 in the denominator of (6.11): Replacing n by $2n$ changes $1/n^4$ to $1/(16n^4)$.

In Examples 6.6, page 162, and 6.7, page 165, we see that the anticipated improvement is actually achieved. For T_n, going from 10 to 20 intervals results in an error one fourth as large. A similar phenomenon occurs in going from 50 to 100 intervals, or from 100 to 200. For S_n, the ratio of the errors for $n = 10$ and $n = 20$ intervals is $1/16$, as expected.

6.15   The error estimates for T_n and S_n are not always valid, for they assume the existence of either 2 or 4 bounded derivatives of the integrand. Indeed, the approximation of the integral of a function which lacks the required number of derivatives may be quite inferior to what we've come to expect, as we'll see in this problem.

For $n = 4$, 8, 16, and 32, compute T_n and S_n for $\int_0^1 \sqrt{x}\,dx$, and the errors in these approximations. (The exact value of the integral is $2/3$ or $.666667$ to 6 decimal places.)

Solution: Using (6.7) and (6.11), we obtain the following table:

n	T_n	S_n	$|E(T_n)|$	$|E(S_n)|$
4	.643283	.656526	.023385	.010140
8	.658130	.663079	.008536	.003587
16	.663581	.665398	.003085	.001268
32	.665559	.666218	.001108	.000448

In our previous examples Simpson's Rule has always yielded much better approximations than the Trapezoidal Rule. In this case, however, the difference between the efficiency of the two methods is minimal. True, the errors in S_n are somewhat smaller than in T_n. But neither of the two methods is doing a very good job. In each case, doubling the number of intervals causes a reduction in the error by a factor of about .35. In the previous problem we showed that if the integrand has enough derivatives, then the error in T_n should go down by a factor of about $1/4 = .25$ when the number of intervals is doubled, and the error in S_n by a factor of about $1/16 = .0625$. So while both methods fail to live up to their potential, Simpson's Rule is especially disappointing in this case.

Well, what went wrong? The problem lies in the function, $\sqrt{x}$, which has no derivative at $x = 0$ (the tangent line is vertical there). Moral: Before you use the error estimates, make sure that they apply to the problem at hand.

6.16   One of the simplest methods for approximating $\int_a^b f(x)\,dx$ is known as the Midpoint Rule, M_n. It is an approximation by rectangles obtained by partitioning $[a, b]$ into n equally spaced intervals by a set $a = x_0, x_1, x_2, \ldots, x_{n-1}, x_n = b$, letting c_i be the *midpoint* of $[x_{i-1}, x_i]$, $i = 1, 2, \ldots n$, and choosing $y_i = f(c_i)$, $i = 1, 2, \ldots$. The approximation is then given by

$$M_n = h(y_1 + y_2 + y_3 + \cdots + y_{n-1} + y_n).$$

Prove that M_n gives the *exact* value if f is any linear function.

Solution: Suppose f is positive on the subinterval, $[x_{i-1}, x_i]$. We'll use an area argument. The approximation there consists of a rectangle of height $f(c_i)$ and width $h = x_i - x_{i-1}$. But if f is linear, then the triangle that lies above the line $y = f(c_i)$ is congruent to the triangle which lies below it (Figure 6-16). Hence, the area of the rectangle (the approximation) equals that under the line (the value of the integral). This result is also valid if f is not positive over the subinterval.

The fact that M_n is exact for linear functions makes the Midpoint Rule competitive with the Trapezoidal Rule.

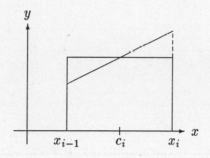

Figure 6-16: The midpoint rule

Supplementary Problems

IMPROPER INTEGRALS

6.17   Evaluate the following improper integrals:

   a. $\displaystyle \int_0^1 \frac{1}{\sqrt[3]{1-x}}\, dx$

   b. $\displaystyle \int_{-\infty}^{\infty} \frac{1}{1+x^2}\, dx$

6.18   Investigate whether the following integrals converge:

   a. $\displaystyle \int_{-\infty}^{\infty} \frac{x^3}{1+x^4}\, dx$

   b. $\displaystyle \int_1^{\infty} \frac{1}{x^2+3x+2}\, dx$

c. $\int_1^\infty \dfrac{\ln x}{x^3}\, dx$

Hint: In the last part use the fact that $\ln x/x < 1$ for all $x \geq 1$.

NUMERICAL INTEGRATION

6.19 Find T_4 and S_4 for the following integrals. (If you have a programmable calcu-lator or a computer, also find T_n and S_n for $n = 8, 16$ and 32.)

a. $\int_0^{\pi/3} x \cos x\, dx$

b. $\int_1^2 \dfrac{e^x}{x}\, dx$

c. $\int_0^1 e^{x^2}\, dx$

6.20 How large should n be to guarantee that T_n is within $.0005$ of $\int_a^b f(x)\, dx$ for each of the following integrals:

a. $\int_0^{\pi/4} \tan x\, dx$

b. $\int_0^1 e^{x^2}\, dx$

6.21 a. If the Trapezoidal Rule is used to approximate $\int_0^1 x^2\, dx$, will T_n be

   - bigger than the integral
   - smaller than the integral
   - impossible to tell?

Hint: Draw a picture.

b. Generalize this result to functions other than x^2. Can you think of a wide class of functions for which a similar result applies?

Answers to Supplementary Problems

6.17 a. 1.5

   b. π

6.18　**a.**　Diverges

　　　b.　Converges

　　　c.　Converges

6.19　**a.**　$T_4 = .398848$, $T_8 = .404890$, $T_{16} = .406397$, $T_{32} = .406774$
$S_4 = .406963$, $S_8 = .406904$, $S_{16} = S_{32} = .406900$, which is correct to 6 decimal places.

　　　b.　$T_4 = 3.068704$, $T_8 = 3.061520$, $T_{16} = 3.059712$, $T_{32} = 3.059267$
$S_4 = 3.059239$, $S_8 = 3.059125$, $S_{16} = S_{32} = 3.059117$, which is correct to 6 decimal places.

　　　c.　$T_4 = 1.490679$, $T_8 = 1.469712$, $T_{16} = 1.464420$, $T_{32} = 1.463094$
$S_4 = 1.463711$, $S_8 = 1.462723$, $S_{16} = 1.462656$, $S_{32} = 1.462652$, which is correct to 6 decimal places.

6.20　**a.**　$n \geq 18$

　　　b.　$n \geq 53$

6.21　**a.**　T_n is bigger than the integral, since the approximating segments all lie *above* the function $f(x) = x^2$.

　　　b.　The same result is true for any function which is concave up.

Chapter 7

Infinite Series

> **What We Know:** How to add *two* numbers.
>
> **What We Want To Know:** How to add an *infinite set* of numbers.
>
> **How We Do It:** We approximate the *infinite sum* with certain *finite sums*.

7.1   Motivation

While calculus is often broken down into two main branches, differential and integral calculus, there is a third area that is of great importance — infinite series. The addition of two or several numbers is the simplest of the arithmetic operations, and is certainly familiar to all of us. We ask the following question: Can we make *any* sense of the notion of adding together the elements of an *infinite* set of numbers? Let's begin with an example which will show that our question is meaningful.

Example 7.1   Consider a square, whose side is 1 and whose area is, hence, also 1. Divide the square into two equal rectangles (Figure 7-1). Then the area of the square is the sum of the areas of these two rectangles, or $1 = 1/2 + 1/2$. Now further divide one

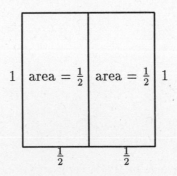

Figure 7-1: Dividing a square into 2 pieces

of the rectangles into two equal squares. The area of the large square is the sum of the areas of the remaining rectangle and the two smaller squares, or $1 = 1/2 + 1/4 + 1/4$ (Figure 7-2). If we again divide one of the smaller squares into two equal rectangles

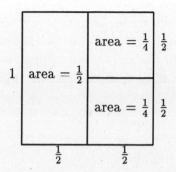

Figure 7-2: Further division of the square

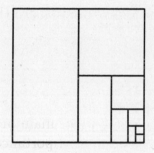

Figure 7-3: Endless division of the square

we will obtain $1 = 1/2 + 1/4 + 1/8 + 1/8$. We now continue the process of subdivision indefinitely (Figure 7-3). The area of the large square is equal to the sum of the areas of all the rectangles and squares thus created. In other words,

$$1 = \frac{1}{2} + \frac{1}{4} + \frac{1}{8} + \frac{1}{16} + \frac{1}{32} + \frac{1}{64} + \cdots.$$

We thus see that the sum of an infinite set of numbers *can be finite*, even though our intuition may have told us otherwise.

This example prompts us to consider the problem in greater generality. Suppose we have an infinite sequence of numbers, which we denote by $a_1, a_2, a_3, \ldots$. (In Example 7.1, above, $a_1 = 1/2$; $a_2 = 1/4 = 1/2^2$; $a_3 = 1/8 = 1/2^3$; and, in general, $a_n = 1/2^n$.) We now ask two questions about any infinite sequence, $a_1, a_2, a_3, \ldots$:

1. Can we determine if there is a finite value which represents the infinite series, $a_1 + a_2 + a_3 + \ldots$?

2. If the answer to our first question is yes, how can we find that value?

Assume that the series has a definite value (as in Example 7.1), but that we don't know what it is. Following our usual approach, we try to *approximate* the value. But

$\mathcal{A}$

when we ask what we can use as an approximation, there is nothing more natural than to turn to the series itself, as follows.

Consider the first term—and it, alone—as an initial approximation (although it is probably not a very good one). To *refine* the approximation, we add the second term in the series to the first; for further refinement, we add the third term to the sum of the first two. We continue in like fashion, obtaining a sequence of approximations that looks like this:

$$a_1$$
$$a_1 + a_2$$
$$a_1 + a_2 + a_3$$
$$\dots\dots\dots\dots\dots\dots$$
$$a_1 + a_2 + a_3 + \cdots + a_n$$

Let's return to Example 7.1. The approximations to the full sum there are:

$$
\begin{aligned}
1/2 &= 1/2 \\
1/2 + 1/4 &= 3/4 \\
1/2 + 1/4 + 1/8 &= 7/8 \\
1/2 + 1/4 + 1/8 + 1/16 &= 15/16 \\
1/2 + 1/4 + 1/8 + 1/16 + 1/32 &= 31/32
\end{aligned}
$$

and, in general,

$$1/2 + 1/4 + 1/8 + 1/16 + \cdots + 1/2^n. \tag{7.1}$$

It appears a pattern is emerging in the successive approximations. The sum in (7.1) can be expressed in *closed form*, namely, it seems to equal $1 - 1/2^n$, since, for example, $31/32 = 1 - 1/32 = 1 - 1/2^5$. Indeed, this is the case, as we will see later on, and this fact is important in evaluating the *full sum*. Assume that we don't know that the sum is 1. How can we find this value?   Up to this point, we have been looking at *approximations* and *refinements*. But we know from our previous work that to obtain the precise or exact value requires us to pass to the *limit*. Now, the sums of the successive approximations are approaching 1, since the nth approximation has sum $1 - 1/2^n$, and $1/2^n$ tends to 0 as n gets larger. Hence the full sum is 1.

Example 7.2 *Determine whether the infinite series*

$$\frac{1}{2} + \frac{1}{6} + \frac{1}{12} + \frac{1}{20} + \frac{1}{30} + \cdots$$

has a finite value and, if so, find that value.

Solution: It is not clear exactly how the series should continue. However, if we notice that $2 = 1 \cdot 2$, $6 = 2 \cdot 3$, $12 = 3 \cdot 4$, $20 = 4 \cdot 5$, and so forth, we see that the general term of the series is

$$a_n = \frac{1}{n(n+1)}.$$

This remark not only clarifies the description of the terms of the series, but, as we will see shortly, will also enable us to evaluate the sum.

Let's play with the series a bit, in the hope that obtaining several approximations will enable us to guess the answer. The successive approximations are:

$$
\begin{aligned}
1/2 &= 1/2 \\
1/2 + 1/6 &= 2/3 \\
1/2 + 1/6 + 1/12 &= 3/4 \\
1/2 + 1/6 + 1/12 + 1/20 &= 4/5 \\
1/2 + 1/6 + 1/12 + 1/20 + 1/30 &= 5/6
\end{aligned}
$$

A pattern is certainly emerging! The denominator of successive terms on the right keeps increasing by one, and each numerator is one less than the corresponding denominator. So an obvious guess for the nth approximation, $1/2 + 1/6 + 1/12 + \cdots + 1/(n(n+1))$, is $n/(n+1)$. If this is true, then the exact value of the series will follow by passing to the limit. Now

$$
\frac{n}{n+1} = 1 - \frac{1}{n+1},
$$

and, since $1/(n+1)$ tends to 0 as n tends to infinity, this limit is 1. But how can we be sure that our guess for the nth approximation is correct? This follows from an alternate expression for the term $1/(n(n+1))$:

$$
\frac{1}{n(n+1)} = \frac{1}{n} - \frac{1}{n+1}
$$

(Although this identity is readily *verified*, it may not be clear just how we obtained the right-hand side from the left. One way of doing so is by use of partial fractions decomposition, an important method for splitting general fractions up into component parts. You can find a discussion of this method in your text in the chapter on Techniques of Integration.)

We use this identity to rewrite the terms of the series, as follows:

$$
\frac{1}{2} + \frac{1}{6} + \frac{1}{12} + \frac{1}{20} + \cdots + \frac{1}{n(n+1)}
$$

$$
= \quad \frac{1}{1 \cdot 2} + \frac{1}{2 \cdot 3} + \frac{1}{3 \cdot 4} + \frac{1}{4 \cdot 5} + \cdots + \frac{1}{n(n+1)}
$$

$$
= \quad \left(1 - \frac{1}{2}\right) + \left(\frac{1}{2} - \frac{1}{3}\right) + \left(\frac{1}{3} - \frac{1}{4}\right) + \left(\frac{1}{4} - \frac{1}{5}\right) + \cdots + \left(\frac{1}{n} - \frac{1}{n+1}\right)
$$

$$
= \quad 1 - \left(\frac{1}{2} - \frac{1}{2}\right) - \left(\frac{1}{3} - \frac{1}{3}\right) - \left(\frac{1}{4} - \frac{1}{4}\right) - \cdots - \left(\frac{1}{n} - \frac{1}{n}\right) - \frac{1}{n+1}
$$

$$
= \quad 1 - \frac{1}{n+1}.
$$

This series is an example of a *telescoping* sum, because of the cancellation of almost all of the terms. As we see, after the cancellations, all we are left with is just $1 - 1/(n+1)$, confirmation of our earlier guess which was based upon numerical observation. So the full sum of the series is, indeed, equal to 1.

Please don't conclude from these examples that all infinite series have sums of 1. In fact, there are series that fail to have definite values at all. An obvious candidate for such a 'bad' series is $1 + 1 + 1 + \cdots$. The successive approximations to this series are $1, 2, 3, 4, \cdots$, with the nth approximation being just n. Because $n \to \infty$, so do the approximations. A more subtle example is the following:

Let $a_1 = 1$, $a_2 = -1$, $a_3 = 1$, $a_4 = -1$ and, in general, the odd terms are all 1, while the even terms are all -1. The successive approximations are clearly $1, 0, 1, 0, 1, 0, \ldots$, and this sequence of numbers does *not* approach a single number as $n \to \infty$. Hence, this series has no limit.

We will see many more examples of both 'good' and 'bad' series, but first let's formalize what we have done and introduce the appropriate notation.

7.2   Definition

The approximating sums that we have been considering in the previous section have a technical name. They are known as the *partial sums* of the *infinite series*. Thus, a_1 is the first partial sum, $a_1 + a_2$ is the second partial sum, and so forth, with the general term, $a_1 + a_2 + \cdots + a_n$ called the nth partial sum. We may think of these partial sums as nothing more than the first, second, ..., nth approximations to the full sum of the series. We denote these partial sums by $s_1, s_2, s_3, \ldots, s_n$. Thus

$$
\begin{aligned}
s_1 &= a_1 \\
s_2 &= a_1 + a_2 \\
s_3 &= a_1 + a_2 + a_3 \\
&\cdots\cdots\cdots\cdots\cdots\cdots\cdots \\
s_n &= a_1 + a_2 + a_3 + \cdots + a_n
\end{aligned}
$$

Definition: We say that the infinite series $a_1 + a_2 + a_3 + \cdots$ has the sum L if the limit as $n \to \infty$ of the sequence of partial sums, $\{s_n\}$, is equal to L. In this case, we also say that the series *converges* to L. If the limit of $\{s_n\}$ does not exist, then we say that the infinite series *diverges*.

Remark 7.1   As in our earlier chapters, we will not enter into the technical aspects of the limit involved here, but, once again, will confine ourselves to an intuitive approach. Consult your textbook for the precise definition.

7.3   Notation

Writing out lengthy sums is cumbersome, so we use the sigma notation introduced in the chapter on the integral. An infinite series is represented in the following form:

$$\sum_{i=1}^{\infty} a_i$$

(read, the sum from 1 to infinity of a_i). The partial sums, s_n, are given by

$$s_n = \sum_{i=1}^{n} a_i.$$

Hence,

$$\sum_{i=1}^{\infty} a_i = \lim_{n \to \infty} s_n,$$

provided the limit exists.

Remark 7.2   Note that we use the letter i as the *index* of summation, and n for the general term in the approximations given by the partial sums, s_n. As a result, the general term of the infinite series is written as a_i (rather than a_n), since n is reserved for the partial sums.

7.4   Computational Techniques

There is an important distinction between the computational techniques in Chapters 2 and 4, and those that we are about to present. The methods we developed earlier allowed us to actually *compute* derivatives and integrals without resorting to their formal definitions, and we might expect similar results here for summing series. Such hopes, however, are dashed by the fact that we can rarely find the *exact* sum of an infinite series. (The reason for this is that changing even one term of an infinity series changes its *sum*, but does not affect the convergence or divergence of the series.) With the exception of one or two classes of series which can be summed explicitly, there are almost no *general* methods for finding the actual sums. As a result, we shift our attention to another part of the definition, the *convergence* or *divergence* of series. Here, we will find many valuable methods which allow us to determine whether or not a series converges. And, once we know that a series converges, we can employ numerical techniques which approximate the sum.

In these days of high-speed computing devices, it is tempting to think that an easy way to determine whether or not a series converges is to find the partial sums, s_n, for large values of n, and see whether they are converging. The following example shows why this approach is problematic.

Example 7.3 *Determine whether the following series (known as the harmonic series) converges or diverges:*

$$\sum_{i=1}^{\infty} \frac{1}{i}.$$

Solution: Let's find s_n for this series for several values of n, small and large:

$$
\begin{aligned}
s_{10} &= 2.928968 \\
s_{100} &= 5.187378 \\
s_{1000} &= 7.485471 \\
s_{10000} &= 9.787606 \\
s_{20000} &= 10.480728 \\
s_{30000} &= 10.886185 \\
s_{40000} &= 11.173863 \\
s_{50000} &= 11.397003
\end{aligned}
$$

If we continue the computation to much larger values of n we will find that the sum of a million terms of the series is less than 15 and even a billion terms add to less than 22. One might think from these calculations that the series converges. But this conclusion is WRONG! For consider the following scheme. Group the terms:

$$
\begin{aligned}
&1 \\
&1/2 \\
&1/3 + 1/4 \\
&1/5 + 1/6 + 1/7 + 1/8 \\
&1/9 + 1/10 + 1/11 + \cdots + 1/16
\end{aligned}
$$

and so forth, and notice the following inequalities:

$$
\begin{aligned}
1 & & &> 1/2 \\
1/2 & & &= 1/2 \\
1/3 + 1/4 & &> 2(1/4) &= 1/2 \\
1/5 + 1/6 + 1/7 + 1/8 & &> 4(1/8) &= 1/2 \\
1/9 + 1/10 + 1/11 + \cdots + 1/16 & &> 8(1/16) &= 1/2
\end{aligned}
$$

Thus

$$
\begin{aligned}
s_1 &> 1/2 &= 1/2 \\
s_2 &> 2(1/2) &= 1 \\
s_4 &> 3(1/2) &= 3/2 \\
s_8 &> 4(1/2) &= 2 \\
s_{16} &> 5(1/2) &= 5/2
\end{aligned}
$$

and, in general,

$$s_{2^n} > (n+1) \cdot \left(\frac{1}{2}\right) = \frac{n+1}{2}.$$

The latter term tends to infinity with increasing n, so that the series *diverges*. We see from this example how dangerous it can be to draw conclusions from numerical work, *without doing previous analysis.*

As mentioned earlier, there are few series which can be summed explicitly. One important exception to this is the geometric series, $\sum_{i=1}^{\infty} u^{i-1}$, whose convergence properties are summarized in the following result:

Theorem 7.1 (Geometric Series)   *The geometric series*

$$\sum_{i=1}^{\infty} u^{i-1} \text{ converges to } \frac{1}{1-u} \text{ if and only if } |u| < 1.$$

For instance,

a. $\displaystyle\sum_{i=1}^{\infty} (.7)^{i-1} = \frac{1}{1-.7} = \frac{1}{.3} = \frac{10}{3}.$

b. $\displaystyle\sum_{i=1}^{\infty} \left(\frac{2}{3}\right)^{i-1} = \frac{1}{1-2/3} = \frac{1}{1/3} = 3.$

c. $\displaystyle\sum_{i=1}^{\infty} \left(-\frac{3}{7}\right)^{i-1} = \frac{1}{1-(-3/7)} = \frac{1}{10/7} = \frac{7}{10}.$

The following is an important test for *divergence*.

Theorem 7.2   *If $\sum_{i=1}^{\infty} a_i$ converges, then $\lim_{i\to\infty} a_i = 0$.*

Proof: Let $s_n = \sum_{i=1}^{n} a_i$, and let L be the limit of the sequence s_n. Now consider $s_n - s_{n-1} = \sum_{i=1}^{n} a_i - \sum_{i=1}^{n-1} a_i = a_n$, since all the other terms cancel. Hence,

$$\lim_{n\to\infty} a_n = \lim_{n\to\infty} (s_n - s_{n-1}) = \lim_{n\to\infty} s_n - \lim_{n\to\infty} s_{n-1} = L - L = 0,$$

and the proof is complete.

The most useful computational techniques are for series which consist solely of positive terms, because in this case there is a simple criterion for convergence. Let's return to the definition of convergence of an infinite series, in which we first compute the partial sums, $s_n = \sum_{i=1}^{n} a_i$. If $a_i > 0$ for all i, then the sequence $\{s_n\}$ is an *increasing* one; that is, $s_1 < s_2 < s_3 < \dots$. Under what circumstances does such a sequence converge? There are only two possibilities for an increasing sequence: Either it grows without bound (that is, $s_n \to \infty$, in which case the series diverges), or else the terms s_n are bounded from above. This means that there exists some number M which is

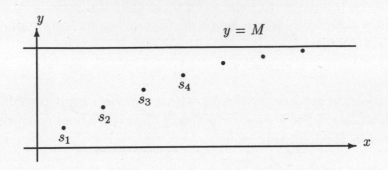

Figure 7-4: An increasing sequence

greater than any element of the sequence: $s_n < M$ for all n. In Example 7.1, $\sum_{i=1}^{\infty} 1/2^i$, the partial sums were, successively, 1/2, 3/4, 7/8, 15/16,..., all bounded from above by 1. The same was true in Example 7.2. Let's look at this geometrically (Figure 7-4). It appears from this diagram that the sequence $\{s_n\}$ is converging to M. However, let us not jump to conclusions. For what is M? It is a number which is greater than every term of the sequence, $\{s_n\}$. In Example 7.1, 5 is also such an *upper bound*, but the sequence does not converge to 5. Nor does it converge to 2.3, 10.5, 117, or 549, although *all* of these numbers are upper bounds of $\{s_n\}$. Among all of the upper bounds (and there are an infinite number of them), what distinguishes the number 1, which *is* the limit of $\{s_n\}$? 1 is called the *least upper bound* of $\{s_n\}$. But what, exactly, do we mean by this? Suppose we have found *an* upper bound, M, for a sequence $\{s_n\}$. Thus M lies above all the terms of the sequence (Figure 7-5). There are two possibilities:

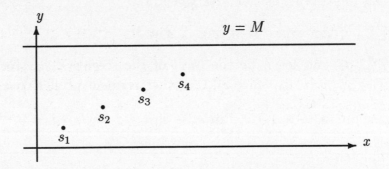

Figure 7-5: M is an upper bound

- We can lower the line $y = M$ and the resulting lowered line is still an upper bound for $\{s_n\}$ (Figure 7-6).

- We cannot lower the line $y = M$ at all without having elements of $\{s_n\}$ exceed M (Figure 7-7).

In the second case, M is the *least upper bound* of $\{s_n\}$. In other words, among *all* upper bounds, it is the *smallest*. If the first possibility holds, however, we *can* lower

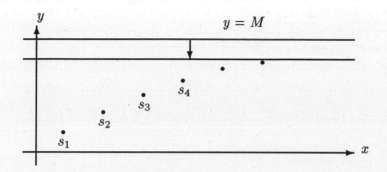

Figure 7-6: M isn't the least upper bound

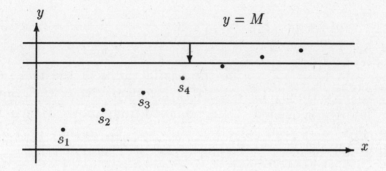

Figure 7-7: M is the least upper bound

the bounding horizontal line. But how far should we lower it? Just far enough so that we reach the second possibility, that is, until we reach the least upper bound. At this point we stop, since we have found the limit of $\{s_n\}$ and, hence, the sum of the series $\sum_{i=1}^{\infty} a_i$.

We conclude our lengthy discussion by summarizing our results to this point. We've shown that if all of the terms a_i are positive, then the sequence of partial sums $\{s_n\}$ is increasing. There are then two possibilities:

- If the sequence $\{s_n\}$ is *bounded*, then $\sum_{i=1}^{\infty} a_i$ *converges*.

- If the sequence $\{s_n\}$ is *unbounded*, then $\sum_{i=1}^{\infty} a_i$ *diverges*.

Remark 7.3 The fact that the lowering process described above always terminates (that is, that the least upper bound always exists), is a deep and important result, known as the completeness property of the real numbers. More details about this topic are found in Advanced Calculus courses.

We restate our result formally.

Theorem 7.3 *If $a_i > 0$ for all i, then $\sum_{i=1}^{\infty} a_i$ converges if and only if the partial sums, s_n, are bounded.*

This theorem is the basis for many of the standard tests for convergence that are found in your textbook, such as the comparison test, limit comparison test, and the integral test. We first illustrate these ideas by introducing the comparison test. In the previous chapter we saw a similar result for improper integrals, so the idea underlying this test is familiar.

Theorem 7.4 (Comparison Test) *Suppose $0 < a_i \leq b_i$ for all i.*

a. *If $\sum_{i=1}^{\infty} b_i$ converges, then so does $\sum_{i=1}^{\infty} a_i$.*

b. *If $\sum_{i=1}^{\infty} a_i$ diverges, then so does $\sum_{i=1}^{\infty} b_i$.*

Proof:

a. Let $s_n = \sum_{i=1}^{n} a_i$ and $t_n = \sum_{i=1}^{n} b_i$ be the partial sums of the two series. Since $a_i \leq b_i$ for all i, we certainly have $s_n \leq t_n$ for all n. Now, the convergence of $\sum_{i=1}^{\infty} b_i$ implies that the sequence $\{t_n\}$ is bounded from above. Hence, there exists a number M such that $t_n \leq M$ for all n. But $s_n \leq t_n$, so that $\{s_n\}$ also satisfies $s_n \leq M$ for all n, and hence, $\sum_{i=1}^{\infty} a_i$ converges, by Theorem 7.3.

Exercise 7.1 *Prove part* **b.** *of Theorem 7.3. (The proof is similar to that of part* **a.***)*

There are many other tests for convergence, all of which are examples of *computational techniques*. Recall, however, as we mentioned at the beginning of this section, that these results tell us only *whether* a series converges, but not the value of its sum. We now state some of these results, and refer you to your text for the proofs. The various tests will be illustrated by examples which, for the most part, are self-contained. As a result, the exposition will be brief.

Tests for series of positive terms:

Theorem 7.5 (Integral Test) *Let f be a continuous, decreasing positive function on the interval $[1, \infty)$ and let $a_i = f(i)$, $i = 1, 2, 3, \ldots$. Then $\sum_{i=1}^{\infty} a_i$ is convergent if and only if the improper integral $\int_1^{\infty} f(x)\, dx$ exists.*

Example 7.4 *Does*

$$\sum_{i=1}^{\infty} \frac{1}{1 + i^2}$$

converge or diverge?

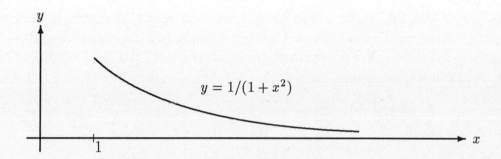

Figure 7-8: Graph of $f(x) = 1/(1 + x^2)$

Solution: Consider the function $f(x) = 1/(1 + x^2)$, which is continuous, decreasing and positive on $[1, \infty)$, with $f(i) = 1/(1 + i^2)$ (Figure 7-8). So the conditions of the Integral Test are fulfilled. We now evaluate the improper integral,

$$\int_1^\infty \frac{1}{1 + x^2}\, dx \quad = \quad \lim_{t \to \infty} \int_1^t \frac{1}{1 + x^2}\, dx \qquad \textit{(definition)}$$

$$= \quad \lim_{t \to \infty} \tan^{-1} x \Big|_1^t \qquad \textit{(Fundamental Theorem)}$$

$$= \quad \lim_{t \to \infty} (\tan^{-1} t - \tan^{-1} 1) \qquad \textit{(substitution)}$$

$$= \quad \lim_{t \to \infty} \tan^{-1} t - \pi/4$$

$$= \quad \pi/2 - \pi/4$$

$$= \quad \pi/4.$$

Since $\int_1^\infty 1/(1 + x^2)\, dx$ exists, the infinite series $\sum_{i=1}^\infty (1/(1 + i^2))$ *converges.*

Example 7.5 *Show that*

$$\sum_{i=1}^\infty \frac{2i}{1 + i^2}$$

diverges.

Solution: Let $f(x) = 2x/(1 + x^2)$, which satisfies all of the conditions needed for the Integral Test. However, in this case, the improper integral does *not* exist, as we see from the following computation.

An antiderivative of $f(x)$ is $\ln(1 + x^2)$. Hence,

$$\int_1^\infty \frac{2x}{1 + x^2}\, dx \quad = \quad \lim_{t \to \infty} \int_1^t \frac{2x}{1 + x^2}\, dx \qquad \textit{(definition)}$$

$$= \quad \lim_{t \to \infty} \ln(1 + x^2) \Big|_1^t \qquad \textit{(Fundamental Theorem)}$$

$$= \lim_{t \to \infty} \ln(1 + t^2) - \ln 2$$

$$= \infty. \qquad \text{(since } \ln x \to \infty \text{ as } x \to \infty)$$

Thus, $\sum_{i=1}^{\infty} 2i/(1 + i^2)$ diverges.

Theorem 7.6 (The p-test) $\sum_{i=1}^{\infty} \dfrac{1}{i^p}$ *converges if and only if $p > 1$.*

This test is a consequence of the Integral Test, together with part (b) of the Theorem on page 157 in Chapter 6. The p-series and the geometric series provide classes of examples which are useful when employing the Comparison Test and the Limit Comparison Test (soon to be stated).

Example 7.6 $\sum_{i=1}^{\infty} \dfrac{1}{i^{1.1}}$ *converges, but* $\sum_{i=1}^{\infty} \dfrac{1}{i^{.9}}$ *diverges.*

Theorem 7.7 (Limit Comparison Test) *Suppose $0 < a_i$ and $0 < b_i$ for all i, and suppose that*

$$\lim_{i \to \infty} \frac{a_i}{b_i} = L,$$

where $0 < L < \infty$. Then either $\sum_{i=1}^{\infty} a_i$ and $\sum_{i=1}^{\infty} b_i$ both converge or else both diverge.

Example 7.7 *Determine whether*

$$\sum_{i=1}^{\infty} \frac{2i + 11}{i^3 - 3i + 4}$$

converges or diverges.

Solution: Let $a_i = (2i + 11)/(i^3 - 3i + 4)$. For i large, the dominant term in the numerator is $2i$, while in the denominator it is i^3. Thus, for i large, a_i seems to 'be near' $2/i^2$.

To make this argument more precise, let $b_i = 1/i^2$. Then

$$\lim_{i \to \infty} \frac{a_i}{b_i} = \lim_{i \to \infty} \frac{(2i + 11)/(i^3 - 3i + 4)}{1/i^2}$$

$$= \lim_{i \to \infty} \frac{i^2(2i + 11)}{i^3 - 3i + 4} \qquad \text{(algebra)}$$

$$= \lim_{i \to \infty} \frac{i^3(2 + 11/i)}{i^3(1 - 3/i^2 + 4/i^3)} \qquad \text{(factoring)}$$

$$= \lim_{i \to \infty} \frac{2 + 11/i}{1 - 3/i^2 + 4/i^3} \qquad \text{(cancelling } i^3\text{)}$$

$$= 2.$$

Now $\sum_{i=1}^{\infty} b_i = \sum_{i=1}^{\infty} 1/i^2$ is a p-series with $p = 2$, and hence is *convergent*. Since $\lim_{i \to \infty} a_i/b_i = 2$, $\sum_{i=1}^{\infty} a_i$ also *converges* by the Limit Comparison Test.

We turn now to series where the terms are not required to be positive. There are far fewer tests in this case, and we present two of the most important ones. The Ratio Test, which we introduce first, will be especially useful in the context of Taylor Series which we'll encounter in the next section.

Tests for general series:

Theorem 7.8 (Ratio Test) *Suppose that the limit of the quotient of the absolute value of successive terms of a series exists. That is, suppose that*

$$\lim_{i \to \infty} \left| \frac{a_{i+1}}{a_i} \right| = L.$$

Then

a. $\sum_{i=1}^{\infty} a_i$ converges *if $L < 1$;*

b. $\sum_{i=1}^{\infty} a_i$ diverges *if $L > 1$;*

c. *if $L = 1$, then the test gives no information about the convergence or divergence of the series.*

Example 7.8 *Investigate the convergence of*

a. $\displaystyle\sum_{i=1}^{\infty} \frac{i}{2^i}$ b. $\displaystyle\sum_{i=1}^{\infty} \frac{i!}{10^i}.$

Solution:

a. Let $a_i = i/2^i$, so that $a_{i+1} = (i+1)/2^{i+1}$. Hence,

$$\begin{aligned}
\lim_{i \to \infty} \frac{a_{i+1}}{a_i} &= \lim_{i \to \infty} \frac{(i+1)/(2^{i+1})}{i/2^i} && \\
&= \lim_{i \to \infty} \left(\frac{i+1}{i} \right) \left(\frac{2^i}{2^{i+1}} \right) && (algebra) \\
&= \lim_{i \to \infty} \left(\frac{i+1}{i} \right) \left(\frac{1}{2} \right) && (cancellation) \\
&= (1/2) \lim_{i \to \infty} (1 + 1/i) && (algebra) \\
&= \frac{1}{2}.
\end{aligned}$$

Since $L = 1/2 < 1$, the series *converges*.

b. Set $a_i = i!/10^i$. Then $a_{i+1} = (i+1)!/10^{i+1}$, so that

$$\frac{a_{i+1}}{a_i} = \frac{(i+1)!/10^{i+1}}{i!/10^i}$$

$$= \left(\frac{(i+1)!}{i!}\right)\left(\frac{10^i}{10^{i+1}}\right) \qquad \text{(algebra)}$$

$$= \frac{i+1}{10}. \qquad \text{(cancellation)}$$

Thus, $\lim_{i\to\infty} a_{i+1}/a_i = \lim_{i\to\infty}(i+1)/10 = \infty$, so that $\sum_{i=1}^{\infty} i!/10^i$ *diverges*.

To show that the Ratio Test is inconclusive when $L = 1$, it suffices to produce two series, one convergent, the other divergent, in both of which $L = 1$. This is easily done, since for each p-series, $L = 1$ (verify), and we know that $\sum_{i=1}^{\infty} 1/i^2$ is convergent, while $\sum_{i=1}^{\infty} 1/i$ diverges.

Theorem 7.9 (Alternating Series Test) *Suppose the following conditions hold:*

a. $a_1 > a_2 > a_3 > \ldots$;

b. $\lim_{i\to\infty} a_i = 0$.

Then the series $\sum_{i=1}^{\infty}(-1)^{i+1}a_i = a_1 - a_2 + a_3 - a_4 + \cdots$ converges.

Example 7.9 *Investigate the convergence of the* alternating harmonic series,

$$1 - \frac{1}{2} + \frac{1}{3} - \frac{1}{4} + \cdots.$$

Solution: The series *converges*, since the terms strictly alternate between positive and negative, and are *decreasing* to 0 in magnitude.

7.5   Application of Series

Perhaps the most important application of infinite series is in the numerical computations of functions. Polynomials and rational functions (quotients of polynomials) can be computed by means of the ordinary arithmetic operations. Other functions require more sophisticated techniques, and infinite series provide one such method. We had a preview of this in Chapter 3, where we used Taylor polynomials as approximations to functions, but at that time, after going through the approximation and refinement stages, we deferred the question of the *limit* of the Taylor process to this chapter, in which we introduce *Taylor series*. We begin our analysis with an example.

Example 7.10 *Find the Taylor series for $\sin x$ about $a = 0$.*

Solution: Let $f(x) = \sin x$. Recall from Section 3.2 that the nth Taylor polynomial at $a = 0$ is given by

$$P_n(x) = f(0) + f'(0)x + \frac{f''(0)x^2}{2!} + \cdots + \frac{f^{(n)}(0)x^n}{n!} \tag{7.2}$$

Here $f'(x) = \cos x$, $f''(x) = -\sin x$, $f'''(x) = -\cos x$ and $f''''(x) = \sin x$, after which the pattern repeats in clusters of 4 derivatives. It is thus sufficient for us to compute $f(x)$, $f'(x)$, $f''(x)$ and $f''''(x)$ at the base point, $a = 0$. Specifically, we have $\sin 0 = 0$ and $\cos 0 = 1$. Hence,

$$
\begin{aligned}
f(0) &= 0 \\
f'(0) &= 1 \\
f''(0) &= 0 \\
f'''(0) &= -1,
\end{aligned}
\tag{7.3}
$$

From (7.2) and (7.3) we find that the *odd* degree Taylor polynomials for $\sin x$ are given by

$$P_{2n-1} = x - \frac{x^3}{3!} + \frac{x^5}{5!} - \frac{x^7}{7!} + \cdots + (-1)^n \frac{x^{2n-1}}{(2n-1)!} \tag{7.4}$$

Note: Since *all* the coefficients of the *even* powers of x are 0, the Taylor polynomial for $\sin x$ of even degree, $2n$, is identical to that of degree $2n - 1$. That is,

$$P_{2n}(x) - P_{2n-1}(x).$$

The error formula for the approximation of $f(x)$ by $P_n(x)$ is

$$E_n(x) = \frac{f^{(n+1)}(c)x^{n+1}}{(n+1)!},$$

for some c between 0 and x. Now $f^{(n+1)}(c)$ is either $\sin c$ or $\cos c$. Regardless of which it is, $|f^{(n+1)}(c)| \le 1$, so that

$$|E_n(x)| \le \frac{|x|^{n+1}}{(n+1)!}, \tag{7.5}$$

and, as was discussed in Remark 3.5 (page 75), for *any* x, the right hand side of (7.5) tends to 0 as n increases. (**Note:** This does *not* mean that the error necessarily gets smaller from the very beginning. In fact, if x is *large*, then the error may actually increase *initially*, for small values of n. For example, $P_3(10) = -156.67$, while $P_5(10) = 676.67$. Since $\sin 10 = -0.54$, both approximations are terrible, but the error in $P_5(10)$ is even worse than that in $P_3(10)$. But no matter how large x is, $E_n(x)$ approaches 0 for n sufficiently large.) Thus, for any x, the approximations are ultimately *refined* by allowing n to increase. For example, consider the approximation to $\sin(\pi/6)$ by $P_8(\pi/6)$. By (7.4) and the note following it, P_8 is identical with P_7. Thus we have

$$P_8(x) = x - \frac{x^3}{3!} + \frac{x^5}{5!} - \frac{x^7}{7!},$$

so that

$$P_8\left(\frac{\pi}{6}\right) = \frac{\pi}{6} - \frac{(\pi/6)^3}{6} + \frac{(\pi/6)^5}{120} - \frac{(\pi/6)^7}{5040}$$

$$= 0.499999992.$$

On the other hand,

$$\sin\left(\frac{\pi}{6}\right) = 0.5,$$

so that

$$\sin\left(\frac{\pi}{6}\right) - P_8\left(\frac{\pi}{6}\right) = 0.000000008,$$

an exceptionally good approximation, obtained by performing nothing more than a few simple arithmetic calculations. If we wish to further improve the approximation, we go on to P_9. This computation is simplified by noticing that

$$P_9(x) = P_8(x) + \frac{f^{(9)}(0)x^9}{9!} = P_8(x) + \frac{x^9}{9!},$$

so that all we need do is to add one more term, $(\pi/6)^9/9!$ to $P_8(\pi/6)$, yielding

$$P_9\left(\frac{\pi}{6}\right) = 0.5000000000$$

(to the degree of accuracy of the calculator; actually, $P_9(\pi/6)$ is *slightly* larger than 0.5, but rounded to 10 decimal places yields 0.5).

Now let's take another value of x, one for which an exact value of the sine isn't known. We approximate $\sin 1$, using P_9 and P_{11}. From (7.4),

$$P_9(1) = 1 - \frac{1}{3!} + \frac{1}{5!} - \frac{1}{7!} + \frac{1}{9!}$$

$$= 0.8414710096$$

and

$$P_{11}(1) = P_9(1) - \frac{1}{11!} = 0.8414709845.$$

If we punch $\sin 1$ directly into a calculator (make sure it's in *radian* mode!), we obtain

$$\sin 1 = 0.8414709848,$$

so that the errors in the approximations by $P_9(1)$ and $P_{11}(1)$ are just 0.0000000151 and 0.0000000003, respectively.

Could we have predicted such a high level of accuracy? Let's turn to the error formula. From (7.5),

$$|E_n(1)| \leq \frac{1}{(n+1)!},$$

so that

$$|E_9(1)| \leq \frac{1}{10!} = 0.0000002758$$

and

$$|E_{11}(1)| \leq \frac{1}{12!} = 0.0000000021.$$

The true errors are much smaller. (Recall that (7.5) gives only a bound on the maximum *possible* error. So we knew in advance from (7.5) that the errors in this case would be very small; we should be pleased that they are even smaller than anticipated.)

At this point we ask whether it makes sense to *take the limit* in (7.4). If we do so, then we'll obtain an *infinite series*,

$$x - \frac{x^3}{3!} + \frac{x^5}{5!} - \frac{x^7}{7!} + \cdots. \tag{7.6}$$

But this series differs from those we have considered previously. Until now, we have dealt with series of *constants*, $\sum_{i=1}^{\infty} a_i$, while here we have a series of *variable* terms. Fixing x, however, puts us back on familiar territory and we can then ask our usual question: Does the series converge? For example, setting $x = 1$ in (7.6), we obtain

$$1 - \frac{1}{3!} + \frac{1}{5!} - \frac{1}{7!} + \cdots.$$

Is this a convergent series? More generally, we can search for *all* the values of x for which (7.6) is convergent. Since, for any specific x, (7.6) is an ordinary series of constants, all of the tests for convergence that we developed in the previous section are applicable. The one that will prove most useful to us is the Ratio Test, which we now restate:

Suppose that

$$\lim_{i \to \infty} \left| \frac{a_{i+1}}{a_i} \right| = L.$$

Then

a. $\sum_{i=1}^{\infty} a_i$ *converges* if $L < 1$;

b. $\sum_{i=1}^{\infty} a_i$ *diverges* if $L > 1$;

c. if $L = 1$, the test gives no information about the convergence or divergence of the series.

In (7.6), the general term, a_i, is

$$\frac{(-1)^{i-1}x^{2i-1}}{(2i-1)!},$$

and the next term, a_{i+1}, is

$$\frac{(-1)^i x^{2i+1}}{(2i+1)!}.$$

Hence,

$$\left|\frac{a_{i+1}}{a_i}\right| = \left|\frac{x^{2i+1}/(2i+1)!}{x^{2i-1}/(2i-1)!}\right|$$

$$= \left|\frac{x^{2i+1}}{x^{2i-1}}\right|\frac{(2i-1)!}{(2i+1)!} \qquad (algebra)$$

$$= \frac{|x^2|}{(2i+1)(2i)}. \qquad (cancellation)$$

Thus,

$$\lim_{i\to\infty}\left|\frac{a_{i+1}}{a_i}\right| = \lim_{i\to\infty}\frac{|x^2|}{(2i+1)(2i)} = 0,$$

no matter what x is. In other words, the series (7.6),

$$x - \frac{x^3}{3!} + \frac{x^5}{5!} - \frac{x^7}{7!} + \cdots$$

is convergent for *every* value of x.

And to what does (7.6) converge? To $\sin x$ (of course!), since, for each x,

$$\lim_{n\to\infty}\left|\sin x - \sum_{i=1}^n \frac{(-1)^{i-1}x^{2i-1}}{(2i-1)!}\right| = \lim_{n\to\infty}|\sin x - P_{2n-1}(x)|$$

$$= \lim_{n\to\infty}|E_{2n-1}(x)|$$

$$= 0.$$

Hence,

$$\sin x = \sum_{i=1}^\infty \frac{(-1)^{i-1}x^{2i-1}}{(2i-1)!}, \qquad (7.7)$$

so that we have an *infinite series representation* of the sine function.

We now turn to a general formulation of this problem. Suppose f has derivatives of *all orders* at $x = a$. The *Taylor series* of f at the point $x = a$ is defined as

$$f(a) + f'(a)(x-a) + \frac{f''(a)(x-a)^2}{2!} + \cdots + \frac{f^{(n)}(a)(x-a)^n}{n!} + \cdots,$$

or

$$\sum_{i=0}^\infty \frac{f^{(i)}(a)(x-a)^i}{i!}. \qquad (7.8)$$

Remark 7.4 To allow for the use of the shorthand notation of (7.8), we define $0! = 1$, and $f^{(0)}(a) = f(a)$.

Remark 7.5 When $a = 0$, (7.8) is also known as the *Maclaurin* series of $f(x)$.

We now ask two questions about (7.8):

1. For which values of x does (7.8) converge?

2. *If* (7.8) *converges for some value of* x, *must it converge to* $f(x)$?

While the answer to the first question is easy to obtain, the second is often harder, as we'll see from the examples that we consider below. Letting $c_i = f^{(i)}(a)/i!$, the series (7.8) takes the form

$$\sum_{i=0}^{\infty} c_i(x - a)^i. \tag{7.9}$$

We now apply the Ratio Test to (7.9), computing

$$
\begin{aligned}
L &= \lim_{i \to \infty} \left| \frac{c_{i+1}(x - a)^{i+1}}{c_i(x - a)^i} \right| \\
&= \lim_{i \to \infty} \left| \frac{c_{i+1}}{c_i} \right| |x - a| \\
&= |x - a| \lim_{i \to \infty} \left| \frac{c_{i+1}}{c_i} \right|. \tag{7.10}
\end{aligned}
$$

Now, (7.9) converges if $L < 1$. So we compute the limit in (7.10), denoting its value by $1/R$. (The reason for this rather strange notation will become apparent shortly.) Thus,

$$
\begin{aligned}
L &= |x - a| \lim_{i \to \infty} \left| \frac{c_{i+1}}{c_i} \right| \\
&= \frac{|x - a|}{R}
\end{aligned}
$$

So (7.9) converges if $L < 1$; that is, if

$$\frac{|x - a|}{R} < 1$$

or

$$|x - a| < R, \tag{7.11}$$

or

$$-R < x - a < R \quad \textit{(expanding the absolute value)}$$

or

$$a - R < x < a + R. \tag{7.12}$$

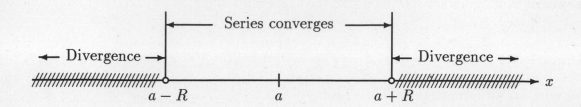

Figure 7-9: The interval of convergence of a Taylor series

R is called the *radius of convergence* of the series (7.8) or (7.9). Note also, by the Ratio Test, that if $L > 1$, then (7.9) *diverges*. Thus (7.9) diverges if $|x - a| > R$ (Figure 7-9).

The only remaining ambiguity is for $L = 1$; that is, when $|x - a| = R$, or $x = a + R$ or $a - R$. In *general*, nothing can be said about the convergence or divergence of the series at these two points, since the series may converge at *both* of these points, at *neither*, or at *just one* of them. We summarize our results.

Theorem 7.10   *Let*

$$R = \lim_{i \to \infty} \left| \frac{c_i}{c_{i+1}} \right|. \tag{7.13}$$

a.   *If $0 < R < \infty$, then $\sum_{i=0}^{\infty} c_i(x - a)^i$ converges for all x satisfying $|x - a| < R$. It may or may not converge at the endpoints, $x = a \pm R$.*

b.   *If $R = \infty$, then the series converges for all x.*

c.   *If $R = 0$, then the series converges only for $x = a$.*

Example 7.11   *The natural logarithm function, $\ln x$, is defined by an integral,*

$$\ln x = \int_1^x \frac{1}{t}\, dt, \quad x > 0.$$

Obtain a series expansion of $\ln x$ which can be used for computational purposes.

Solution: We can't expand $\ln x$ about $a = 0$, since $\ln x$ is defined only for $x > 0$. We could expand it about $a = 1$, but a slightly more convenient approach is to expand the function $f(x) = \ln(1 + x)$ about $a = 0$, which we now proceed to do.

We set up the computation as follows:

$$
\begin{array}{llll}
f(x) & = \ln(1+x) & f(0) & = 0 \\
f'(x) & = 1/(1+x) & f'(0) & = 1 \\
f''(x) & = -1/(1+x)^2 & f''(0) & = -1 \\
f'''(x) & = 2/(1+x)^3 & f'''(0) & = 2 \\
f^{(iv)}(x) & = -3!/(1+x)^4 & f^{(iv)}(0) & = -3! \\
f^{(v)}(x) & = 4!/(1+x)^5 & f^{(v)}(0) & = 4!
\end{array}
$$

In general,

$$f^{(i)}(x) = \frac{(-1)^{i-1}(i-1)!}{(1+x)^i},$$

so that

$$f^{(i)}(0) = (-1)^{i-1}(i-1)!. \tag{7.14}$$

Hence,

$$c_i = \frac{f^{(i)}(0)}{i!} = \frac{(-1)^{i-1}(i-1)!}{i!} = \frac{(-1)^{i-1}}{i}.$$

So the Taylor series for $f(x) = \ln(1+x)$ becomes

$$x - \frac{x^2}{2} + \frac{x^3}{3} - \frac{x^4}{4} + \frac{x^5}{5} - \cdots = \sum_{i=1}^{\infty} \frac{(-1)^i}{i} x^i. \tag{7.15}$$

We now compute the radius of convergence. From (7.13)

$$\begin{aligned}
R &= \lim_{i \to \infty} \left| \frac{c_i}{c_{i+1}} \right| \\
&= \lim_{i \to \infty} \frac{1/i}{1/(i+1)} \\
&= \lim_{i \to \infty} \frac{i+1}{i} = 1.
\end{aligned}$$

Hence, $R = 1$, so that (7.15) *converges* in $(-1, 1)$ and *diverges* for $x < -1$ and $x > 1$. Now for the endpoints:

- At $x = 1$, (7.15) becomes the alternating harmonic series,

$$1 - \frac{1}{2} + \frac{1}{3} - \frac{1}{4} + \cdots,$$

which we saw earlier (Example 7.9) *converges*.

- At $x = -1$, (7.15) becomes

$$-1 - \frac{1}{2} - \frac{1}{3} - \frac{1}{4} - \cdots = -\left(1 + \frac{1}{2} + \frac{1}{3} + \frac{1}{4} + \cdots\right),$$

and the series in parentheses is the *divergent* harmonic series (Example 7.3).

Putting together all of the above, we see that the exact *interval of convergence* of (7.15) is $(-1, 1]$ (Figure 7-10).

Fine. We now know *where* the series (7.15) converges. But to *what* does it converge? We *hope* that it converges to $\ln(1+x)$, but we have no assurance of this. To answer this question, we must analyze the error in the approximation of $\ln(1+x)$ by Taylor *polynomials*, $P_n(x)$, and see if this error tends to 0 with increasing n.

Figure 7-10: The interval of convergence is $(-1, 1]$

From (3.23) on page 74, we know that a formula for the error $E_n(x) = f(x) - P_n(x)$ is given by

$$E_n(x) = \frac{f^{(n+1)}(c)(x-a)^{n+1}}{(n+1)!}, \tag{7.16}$$

for some c between a and x. From (7.14) we have

$$f^{(n+1)}(x) = \frac{(-1)^n n!}{(1+x)^{n+1}},$$

so that (7.16) becomes

$$E_n(x) = \frac{(-1)^n}{n+1} \left(\frac{x}{1+c}\right)^{n+1}, \tag{7.17}$$

where c is between 0 and x. For $-1/2 < x \le 1$,

$$\left|\frac{x}{1+c}\right| < 1,$$

so that $|E_n(x)|$, which satisfies

$$|E_n(x)| \le \left(\frac{x}{1+c}\right)^{n+1} \frac{1}{(n+1)},$$

tends to 0 as $n \to \infty$. Hence, the series (7.15) converges to $\ln(1+x)$ for $-1/2 < x \le 1$.

However, for $-1 < x \le -1/2$, we *cannot* show from (7.17) that $E_n(x) \to 0$, since, in this case,

$$\left|\frac{x}{1+c}\right|$$

could be greater than 1. However, the problem lies in the *form* of the error (7.16), which is too weak (in this case) to provide the estimate we need, even though the result is true. The proof relies upon an alternative, integral form of the error, $E_n(x)$, which we haven't considered. Nevertheless, we will *assume* the truth of this result and conclude that the series (7.15) represents the function $\ln(1+x)$ throughout $(-1, 1]$.

Let's put aside the question of the exact interval of convergence, and return to the original purpose of Example 7.11, which was the numerical calculation of logarithms. We *have* shown, legitimately, that the series (7.15) does converge to $\ln(1+x)$ for $-1/2 < x \le 1$, so that we *can* use this series to compute the natural logarithm of any number between 1/2 and 2. Amazingly, this is all we need to know in order to compute the logarithm of *any* positive number! This fact follows from some basic properties of logarithms:

a. $\ln(a \cdot b) = \ln a + \ln b$;

b. $\ln a/b = \ln a - \ln b$;

c. $\ln a^b = b \ln a$;

d. $\ln(1/a) = -\ln a$.

For example, let's compute $\ln 3$. Using the properties of logs mentioned above, we obtain

$$\ln 3 = \ln\left[2\left(\frac{3}{2}\right)\right] = \ln 2 + \ln 1.5,$$

and we know (or can compute) these values from the Taylor series for $\ln(1+x)$.

Similarly, let's find $\ln 37$. Now every number lies between two successive powers of 2; in this case, $32 < 37 < 64$ ($32 = 2^5$ and $64 = 2^6$). So we write

$$37 = 32\left(\frac{37}{32}\right).$$

Hence,

$$
\begin{aligned}
\ln 37 &= \ln\left[32\left(\tfrac{37}{32}\right)\right] \\[2mm]
&= \ln 32 + \ln \tfrac{37}{32} && \textit{(by \textbf{a}.)} \\[2mm]
&= \ln 2^5 + \ln 1.15625 && \textit{(arithmetic)} \\[2mm]
&= 5\ln 2 + \ln 1.15625 && \textit{(by \textbf{c}.),}
\end{aligned}
$$

and we can compute *both* $\ln 2$ and $\ln 1.15625$ from the series (7.15). In this way, we can find the logarithm of any number $x > 1$. Just locate x between two successive powers of 2 and proceed as above.

For $0 < x \leq 1/2$, we use property (d). For example,

$$
\begin{aligned}
\ln 0.374 &= -\ln\left(\frac{1}{0.374}\right) \\[3mm]
&= -\ln 2.6738,
\end{aligned}
$$

which we can find from (7.15).

In all honesty we should point out that (7.15) is actually an inefficient method for calculating logarithms, since the series converges very slowly. However, modifications of the ideas we've introduced *do* provide efficient approximations to $\ln x$.

Example 7.12 *Expand $f(x) = e^x$ in a Taylor series about $a = 0$.*

Solution: Here, all of the derivatives are also equal to e^x, so that $f^{(i)}(0) = e^0 = 1$, for all i. Hence, the expansion of e^x is given by

$$1 + x + \frac{x^2}{2!} + \frac{x^3}{3!} + \cdots = \sum_{i=0}^{\infty} \frac{x^i}{i!}. \tag{7.18}$$

In this case, it is easy to show that the series converges *everywhere* to e^x. The coefficients are $c_i = 1/i!$, so that (7.13) yields

$$
\begin{aligned}
R &= \lim_{i \to \infty} \left| \frac{c_i}{c_{i+1}} \right| \\[2mm]
&= \lim_{i \to \infty} \frac{1/i!}{1/(i+1)!} \\[2mm]
&= \lim_{i \to \infty} \frac{(i+1)!}{i!} \\[2mm]
&= \lim_{i \to \infty} (i+1) = \infty.
\end{aligned}
$$

Hence, the series (7.18) converges for all x. To show that the limit is e^x, we investigate the error estimate (7.16), which *is* adequate in this case to yield the desired result. Since $f^{(n+1)}(x) = e^x$, we have

$$E_n(x) = \frac{e^c x^{n+1}}{(n+1)!},$$

for some c between 0 and x. Now,

$$\lim_{n \to \infty} \frac{|x|^{n+1}}{(n+1)!} = 0$$

for any x, so that the error tends to 0. Hence,

$$e^x = 1 + x + \frac{x^2}{2!} + \frac{x^3}{3!} + \cdots.$$

Solved Problems

7.1   Find the limits of the following sequences:

    a. $\displaystyle \lim_{n \to \infty} \frac{n+1}{n}$

    b. $\displaystyle \lim_{n \to \infty} \frac{\sin n}{n}$

Solution:

a. $\dfrac{n+1}{n} = 1 + \dfrac{1}{n}$ and $\lim\limits_{n\to\infty} \dfrac{1}{n} = 0$, so the limit is 1.

b. $\left|\dfrac{\sin n}{n}\right| \leq \dfrac{1}{n}$, which tends to 0 as $n \to \infty$, so the limit is 0.

7.2 An important result in the theory of the convergence of sequences is known as the 'Sandwich Theorem,' which states the following:

Suppose

$$a_n \leq b_n \leq c_n, \quad n = 1, 2, \ldots,$$

and

$$\lim_{n\to\infty} a_n = \lim_{n\to\infty} c_n = L.$$

Then

$$\lim_{n\to\infty} b_n = L.$$

Here, the sequence $\{b_n\}$ is 'sandwiched' between $\{a_n\}$ and $\{c_n\}$. Since both $\{a_n\}$ and $\{c_n\}$ are approaching L, so is $\{b_n\}$.

Use the Sandwich Theorem to prove that the sequence $b_n = 2^n/n!$ converges to 0.

Solution:

$$0 \leq \frac{2^n}{n!} = \left(\frac{2}{n}\right)\left(\frac{2}{n-1}\right)\left(\frac{2}{n-2}\right)\cdots\left(\frac{2}{4}\right)\left(\frac{2}{3}\right)\left(\frac{2}{2}\right)\left(\frac{2}{1}\right) \leq \left(\frac{8}{6}\right)\left(\frac{2}{4}\right)^{n-3}.$$

Choose $a_n = 0$ (the sequence which is constantly 0), and

$$c_n = \left(\frac{4}{3}\right)\left(\frac{1}{2}\right)^{n-3}.$$

Then $a_n \leq b_n \leq c_n$ and $\lim_{n\to\infty} c_n = 0$. Thus, by the Sandwich Theorem, $\{b_n\}$ also converges to 0.

7.3 Find s_3 and s_6 (the third and sixth partial sums) for the series $\sum\limits_{i=1}^{\infty} \dfrac{2i-1}{2^i}$.

Solution: Write out the partial sums:

$$s_3 = \frac{1}{2} + \frac{3}{4} + \frac{5}{8} = \frac{15}{8}$$

and

$$s_6 = s_3 + \frac{7}{16} + \frac{9}{32} + \frac{11}{64} = \frac{177}{64}.$$

7.4   Test the following series for convergence. Where possible, find the sum of the series.

a. $\displaystyle\sum_{i=2}^{\infty} \frac{1}{i\ln i}$ b. $\displaystyle\sum_{i=2}^{\infty} \frac{(-1)^i}{i\ln i}$ c. $\displaystyle\sum_{i=0}^{\infty} \frac{(-1)^i}{4^i}$ d. $\displaystyle\sum_{i=1}^{\infty} \frac{i}{i^2+1}$

e. $\displaystyle\sum_{i=1}^{\infty} \frac{i^4-2i+4}{2i^4+1}$ f. $\displaystyle\sum_{i=1}^{\infty} \frac{2^i i!}{(2i)!}$

Solution:

a.   The function $f(x) = 1/(x\ln x)$ is positive and decreasing on the interval $[2, \infty)$. Hence, by the Integral Test (page 188), the series converges if and only if the improper integral

$$\int_2^{\infty} \frac{dx}{x\ln x}$$

converges. So we evaluate the associated integral.

$$\int_2^{\infty} \frac{dx}{x\ln x} \;=\; \lim_{t\to\infty} \int_2^t \frac{dx}{x\ln x}$$

$$=\; \lim_{t\to\infty} \ln(\ln x)\Big|_2^t$$

$$=\; \lim_{t\to\infty} \ln(\ln t) - \ln(\ln 2),$$

which tends to ∞. Hence, the integral *diverges* and so does the series.

b.   We saw in the previous part that $1/(i\ln i)$ decreases to 0. Since the terms of the series are alternating in sign, we can apply the Alternating Series Test (page 192), which shows that the series *converges*.

c.   This is a geometric series with $u = -1/4$. Thus, the series *converges* to $1/(1-u) = 4/5$.

d.   Apply the Limit Comparison Test (page 190) using the companion series $\sum_{i=1}^{\infty} 1/i$. Since this series diverges and

$$\lim_{n\to\infty} \frac{i/(i^2+1)}{1/i} = 1,$$

the Limit Comparison Test tells us that the given series also *diverges*.

e.   By Theorem 7.2 (page 185), convergence of a series is possible only if its terms tend to 0. In this case, the terms of the series tend to $1/2$, so the series *diverges*.

f.   Use the Ratio Test (page 191). Letting $a_i = 2^i i!/(2i)!$, we obtain $a_{i+1} = 2^{i+1}(i+1)!/(2i+2)!$, so that

$$
\begin{aligned}
L &= \lim_{i \to \infty} \frac{a_{i+1}}{a_i} \\
&= \lim_{i \to \infty} \frac{2^{i+1}(i+1)!/(2i+2)!}{2^i i!/(2i)!} \\
&= \lim_{i \to \infty} \frac{2(i+1)}{(2i+2)(2i+1)} \\
&= \lim_{i \to \infty} \frac{1}{2i+1} = 0.
\end{aligned}
$$

Since $L < 1$ the series *converges*.

7.5   Test for convergence:

a. $2 - 3/2 + 4/3 - 5/4 + 6/5 - 7/6 + \cdots$

b. $\dfrac{1}{3 \cdot 5} + \dfrac{1}{5 \cdot 7} + \dfrac{1}{7 \cdot 9} + \dfrac{1}{9 \cdot 11} + \cdots$

Solution:

a.   At first sight, this looks like an alternating series. The terms are strictly alternating in sign and they are getting smaller. However, *they are not tending to 0* and, by Theorem 7.2, this fact automatically means that the series *diverges*.

b.

$$
\frac{1}{3 \cdot 5} + \frac{1}{5 \cdot 7} + \frac{1}{7 \cdot 9} + \cdots \quad < \quad \frac{1}{3^2} + \frac{1}{5^2} + \frac{1}{7^2} + \cdots
$$

$$
< \quad \frac{1}{1^2} + \frac{1}{2^2} + \frac{1}{3^2} + \cdots,
$$

which converges (it's a p-series, with $p = 2$). By the Comparison Test, our series also *converges*.

7.6   Prove that the series

$$
\sum_{i=1}^{\infty} \frac{1}{4i - 1} = \frac{1}{3} + \frac{1}{7} + \frac{1}{11} + \frac{1}{15} + \cdots
$$

diverges by means of the

a.   Integral Test

b.   Comparison Test

c.   Limit Comparison Test.

Will the Ratio Test work for this series?

Solution:

a.   Consider the improper integral

$$\int_1^\infty \frac{dx}{4x-1} = \lim_{t\to\infty} \int_1^t \frac{dx}{4x-1}$$

$$= \lim_{t\to\infty} \frac{1}{4} \ln(4x-1)\Big|_1^t$$

$$= \frac{1}{4}\left(\lim_{t\to\infty} \ln(4t-1) - \ln 3\right),$$

which does not exist. So the series *diverges*.

b.

$$\frac{1}{3}+\frac{1}{7}+\frac{1}{11}+\frac{1}{15}+\cdots \quad > \quad \frac{1}{4}+\frac{1}{8}+\frac{1}{12}+\frac{1}{16}+\cdots$$

$$= \frac{1}{4}\left[\frac{1}{1}+\frac{1}{2}+\frac{1}{3}+\frac{1}{4}+\cdots\right].$$

So the given series is greater than 1/4 of the divergent harmonic series. By the Comparison Test, it, too, *diverges*.

c.   Compare the series with the divergent harmonic series, $\sum_{i=1}^\infty 1/i$. Since

$$\lim_{i\to\infty} \frac{1/(4i-1)}{1/i} = \frac{1}{4},$$

the Limit Comparison Test tells us that our series *diverges*.

The Ratio Test does not yield any information, since

$$\lim_{i\to\infty} \frac{a_{i+1}}{a_i} = \lim_{i\to\infty} \frac{1/(4i+3)}{1/(4i-1)} = 1.$$

7.7   Prove that if $a_i > 0$ for all i and if $\sum_{i=1}^\infty a_i$ converges, then $\sum_{i=1}^\infty a_i^2$ also converges.

Solution: Here's an informal proof: Since $\sum_{i=1}^\infty a_i$ converges, $\lim_{i\to\infty} a_i = 0$ from Theorem 7.2. So, from some point on, all of the terms of the series must be smaller than one. However, the square of a number less than 1 is smaller than the number. So if $a_i < 1$, then $a_i^2 < a_i$, and we can use the Comparison Test to show convergence.

There's a slight problem with this proof: The Comparison Test requires that *every* term of the series be smaller than that of the companion series, and here we know only that $a_i^2 < a_i$ *from some point on.* However, there's a way out, and it allows us to emphasize an important point. Convergence or divergence of a series is completely dependent on the "tail" of the series. Changing the first 1000 terms of a convergent series does not affect its convergence. (It will, of course, affect the *sum* of the series.) Similarly, you cannot convert a divergent series into a convergent one by changing a *finite* number of terms. As a result of this discussion, the Comparison Test is valid if the inequality $0 < a_i \leq b_i$ holds from some point on, and this suffices to prove our result.

7.8   A nice application of geometric series is to infinite decimals. We're familiar with ordinary decimals, but what do we mean by an infinite decimal? Well, suppose you divide 1 by 3 (long division). The result is a never-ending string of 3s:

$$\frac{1}{3} = .3333\ldots.$$

Do the same with 1/11 and you get .090909.... A similar thing happens with $1/7 = .142857142857\ldots.$ Notice that, in all of these cases, the decimal we obtain is *repeating.* In fact, whenever we convert a fraction to a decimal, one of two things will happen: Either the division terminates (for example, $1/4 = .25$ or $9/5 = 1.8$), or else a repeating infinite decimal is obtained. The purpose of this problem is to investigate the converse of this result. We will show that any infinite decimal is a fraction.

Find the fractional equivalent of

a.   .373737....

b.   2.4637637637....

Solution:

a.

$$.373737\ldots = .37 + .0037 + .000037 + \cdots$$
$$= .37\left[1 + .01 + .0001 + \cdots\right]$$
$$= \frac{37}{100}\left[1 + .01 + (.01)^2 + (.01)^3 + \cdots\right].$$

Now, the infinite series in brackets is a geometric one with $u = .01$. Hence, its sum is $1/(1 - u) = 1/(.99) = 100/99$, so that

$$.373737\ldots = \left(\frac{37}{100}\right)\left(\frac{100}{99}\right) = \frac{37}{99}.$$

b.

$$2.4637637637\ldots \quad = \quad 2.4 + .0637 + .0000637 + \cdots$$

$$= \quad 2.4 + .0637\,[1 + .001 + .000001 + \cdots]$$

$$= \quad 2.4 + \frac{637}{10000}\,[1 + .001 + (.001)^2 + (.001)^3 + \cdots]$$

$$= \quad \frac{24}{10} + \frac{637}{10000} \cdot \frac{1000}{999}$$

$$= \quad \frac{24}{10} + \frac{637}{9990}$$

$$= \quad \frac{24613}{9990}.$$

7.9 Compute the Taylor series for $f(x) = \cos x$ around $a = 0$, and find its radius of convergence.

Solution: We begin by computing the derivatives of f at 0.

$$
\begin{array}{ll}
f(x) = \cos x & f(0) = 1 \\
f'(x) = -\sin x & f'(0) = 0 \\
f''(x) = -\cos x & f''(0) = -1 \\
f'''(x) = \sin x & f'''(0) = 0 \\
f''''(x) = \cos x & f''''(0) = 1
\end{array}
$$

From this point on the derivatives repeat in groups of 4. Thus, the Taylor series of $\cos x$ is

$$1 - \frac{x^2}{2!} + \frac{x^4}{4!} - \frac{x^6}{6!} + \cdots.$$

From (7.13), to find the radius of convergence we evaluate

$$R = \lim_{i \to \infty} \left| \frac{c_i}{c_{i+1}} \right| = \lim_{i \to \infty} \frac{1/(2i)!}{1/(2i+2)!} = \lim_{i \to \infty} (2i+2)(2i+1) = \infty.$$

So the series converges for all x.

7.10 Differentiate the Taylor series of $\sin x$ term-by-term and show that the resulting series is the Taylor series for $\cos x$.

Solution: The series for $\sin x$ is

$$x - \frac{x^3}{3!} + \frac{x^5}{5!} - \frac{x^7}{7!} + \cdots.$$

Differentiating term-by-term yields

$$1 - 3 \cdot \frac{x^2}{3!} + 5 \cdot \frac{x^4}{5!} - 7 \cdot \frac{x^6}{7!} = 1 - \frac{x^2}{2!} + \frac{x^4}{4!} - \frac{x^6}{6!} + \cdots,$$

which we have just seen is the Taylor series for $\cos x$.

7.11   From the previous problem, it appears that Taylor series may adhere to the rules of calculus. Indeed, this is generally the case, although there are some theoretical considerations here which we won't enter into. They also satisfy rules of algebra. For example, we know that the sum of the geometric series $\sum_{i=0}^{\infty} x^i$ is $1/(1-x)$, for $|x| < 1$. We can obtain many other series from this one by *substitution*.

Find Taylor series for

   a. $\dfrac{1}{1+x}$ **b.** $\dfrac{1}{1+x^2}$ **c.** $\tan^{-1} x$

Solution:

a.   To find the series for $1/(1+x)$ simply replace x by $-x$ in the series $1 + x + x^2 + x^3 + \cdots$. We obtain $1 - x + x^2 - x^3 + \cdots$ as the Taylor series for $1/(1+x)$. The series expansion is valid for $|x| < 1$.

b.   Just replace x by x^2 in the last series. Thus

$$\frac{1}{1+x^2} = 1 - x^2 + x^4 - x^6 + \cdots, \quad |x| < 1.$$

c.   The series for $\tan^{-1} x$ is obtained by *integrating* the last series term-by-term:

$$\tan^{-1} x = x - \frac{x^3}{3} + \frac{x^5}{5} - \frac{x^7}{7} + \cdots.$$

This is valid in $|x| \leq 1$. The validity at the endpoints, $x = \pm 1$, follows because

$$1 - \frac{1}{3} + \frac{1}{5} - \frac{1}{7} + \cdots$$

is an alternating series.

7.12   **Always-sometimes-never:**

   a.   If $\sum_{i=0}^{\infty} c_i(x-2)^i$ converges at $x = 4$, then it _____ converges at $x = 1$.
   b.   If $\sum_{i=0}^{\infty} c_i(x+5)^i$ diverges at $x = 1$, then it _____ converges at $x = -12$.
   c.   If $\sum_{i=0}^{\infty} c_i(x-1)^i$ converges at $x = 0$, then it _____ converges at $x = 2$.

Solution:

a.   Since $\sum_{i=0}^{\infty} c_i(x-2)^i$ converges at $x = 4$, its radius of convergence is *at least* 2, so that the interval of convergence contains the interval $(0, 4]$. Since 1 belongs to this interval, the series *always* converges at $x = 1$.

b.   Since $\sum_{i=0}^{\infty} c_i(x+5)^i$ diverges at $x = 1$, its radius of convergence is *at most* 6. Thus, the largest possible interval of convergence extends from $x = -11$ to $x = 1$. Since -12 falls outside this interval, the series *never* converges at $x = -12$.

c.   Since $\sum_{i=0}^{\infty} c_i(x-1)^i$ converges at $x = 0$, its radius of convergence is at least 1. If it is greater than 1, then the series will converge at $x = 2$, whose distance from the point of expansion is 1. However, the radius of convergence may be *exactly* 1. In this case, $x = 2$ would be an endpoint of the interval of convergence, and we've seen that the series may or may not converge at an endpoint. Hence, the series *sometimes* converges at $x = 2$.

7.13   If r is a positive integer, then the binomial $(1+x)^r$ may be computed by algebraically multiplying out the terms, obtaining familiar expressions such as $(1+x)^2 = 1 + 2x + x^2$ or $(1+x)^3 = 1 + 3x + 3x^2 + x^3$. However, if r is not a positive integer, then the expansion of the binomial results in an infinite series, as we'll see in this problem.

Compute the Maclaurin series for $\sqrt{1+x} = (1+x)^{1/2}$

Solution: Let $f(x) = (1+x)^{1/2}$. Then

$$f(x) = (1+x)^{1/2} \qquad\qquad f(0) = 1$$

$$f'(x) = \tfrac{1}{2}(1+x)^{-1/2} \qquad\qquad f'(0) = 1/2$$

$$f''(x) = (\tfrac{1}{2})(-\tfrac{1}{2})(1+x)^{-3/2} \qquad\qquad f''(0) = -1/2^2$$

$$f'''(x) = (\tfrac{1}{2})(-\tfrac{1}{2})(-\tfrac{3}{2})(1+x)^{-5/2} \qquad f'''(0) = 3/2^3$$

Continuing in like fashion, we find that the general derivative is

$$f^{(i)}(0) = (-1)^{i-1}\frac{1 \cdot 3 \cdot 5 \cdots (2i-3)}{2^i},$$

so that the coefficient of the general term of the series $\sum_{i=0}^{\infty} c_i x^i$ is

$$c_i = (-1)^{i-1}\frac{1 \cdot 3 \cdot 5 \cdots (2i-3)}{2^i \cdot i!}.$$

To find the radius of convergence, we compute

$$R = \lim_{i \to \infty}\left|\frac{c_i}{c_{i+1}}\right|$$

$$= \lim_{i \to \infty}\left|\frac{[1 \cdot 3 \cdot 5 \cdots (2i-3)]/[2^i \cdot i!]}{[1 \cdot 3 \cdot 5 \cdots (2i-3)(2i-1)]/[2^{i+1} \cdot (i+1)!]}\right|$$

$$= \lim_{i \to \infty}\frac{2i+2}{2i-1} = 1.$$

Hence, the radius of convergence is 1.

7.14   In Solved Problem 6.10 we estimated $\int_0^1 e^{-x^2}\,dx$ using Simpson's Rule with 4 subintervals. Let's see now how well series do in approximating this integral.

Compute the Maclaurin expansion for e^{-x^2} and use four terms to approximate $\int_0^1 e^{-x^2}\,dx$.

Solution: Begin with the Maclaurin series for e^x:

$$e^x = 1 + x + \frac{x^2}{2!} + \frac{x^3}{3!} + \cdots .$$

Replacing x by $-x^2$, we obtain

$$e^{-x^2} = 1 - x^2 + \frac{x^4}{2!} - \frac{x^6}{3!} + \cdots .$$

Now integrate the first four terms on the right-hand side from 0 to 1:

$$\int_0^1 \left(1 - x^2 + \frac{x^4}{2!} - \frac{x^6}{3!} \right) dx = \left(x - \frac{x^3}{3} + \frac{x^5}{10} - \frac{x^7}{42} \right)\Big|_0^1 = \frac{26}{35} = .742857.$$

In Solved Problem 6.10 we obtained the value of .746855 for S_4. The true value (to 6 decimal places) is .746824, so Simpson's Rule is much more accurate in this case.

Supplementary Problems

7.15   **a.**   Write out the first 6 terms of the sequence

$$a_n = \frac{n-1}{n}.$$

Does the sequence converge?

b.   Answer the same questions for the sequence

$$b_n = (-1)^{n-1}\frac{n-1}{n}.$$

7.16   What is a general formula for the sequence

$$1, \frac{1}{\sqrt{2}}, \frac{1}{\sqrt{3}}, \frac{1}{2}, \ldots ?$$

7.17   Test the following series for convergence:

a. $\dfrac{1}{\sqrt{1}} + \dfrac{1}{\sqrt{2}} + \dfrac{1}{\sqrt{3}} + \dfrac{1}{\sqrt{4}} + \cdots.$

b. $\dfrac{1}{\sqrt{1}} - \dfrac{1}{\sqrt{2}} + \dfrac{1}{\sqrt{3}} - \dfrac{1}{\sqrt{4}} + \cdots.$

c. $\displaystyle\sum_{i=0}^{\infty} \left[\left(\dfrac{1}{3} \right)^i - \left(\dfrac{3}{4} \right)^i \right]$

d. $\displaystyle\sum_{i=0}^{\infty} \dfrac{1}{i^2 + i + 1}$

7.18   Find a rational number which is equal to

a. $.16666\ldots$

b. $3.141414\ldots$

c. $.99999\ldots$

7.19   Find Taylor series expansions of the following functions about the indicated point:

a. $\sin x, \quad a = \pi/2$

b. $1/\sqrt{1-x}, \quad a = 0$

c. $e^{2x}, \quad a = 1$

d. $\cos\left(x + \dfrac{\pi}{4} \right), \quad a = 0$

Answers to Supplementary Problems

7.15   a. $0, \dfrac{1}{2}, \dfrac{2}{3}, \dfrac{3}{4}, \dfrac{4}{5}, \dfrac{5}{6}.$ Sequence converges to 1.

b. $0, -\dfrac{1}{2}, \dfrac{2}{3}, -\dfrac{3}{4}, \dfrac{4}{5}, -\dfrac{5}{6}.$ Sequence diverges.

7.16 $\dfrac{1}{\sqrt{n}}.$

7.17   a.   Diverges (It's a p-series, with $p = 1/2$.)

b.   Converges (By the Alternating Series test.)

c. Converges to $-5/2$ (It's the difference of 2 geometric series.)

d. Converges (By comparison with $\sum_{i=0}^{\infty} \frac{1}{i^2}$.)

7.18 a. $1/6$

b. $311/99$

c. 1

7.19 a. $\sum_{i=0}^{\infty}(-1)^i \frac{(x-\pi/2)^{2i}}{(2i)!}$

b. $\sum_{i=0}^{\infty} c_i x^i$, where $c_i = \frac{1 \cdot 3 \cdot 5 \cdots (2i-1)}{2^i \cdot i!}$

c. $e^2 \sum_{i=0}^{\infty} \frac{2^i(x-1)^i}{i!}$

d. $\frac{\sqrt{2}}{2}\left[1 - x - \frac{x^2}{2!} + \frac{x^3}{3!} + \frac{x^4}{4!} - \frac{x^5}{5!} - \frac{x^6}{6!} + \cdots\right]$

Index

TONY JOHNSTON

The Wagon

PAINTINGS BY
JAMES E. RANSOME

Tambourine Books ❖ New York

Printed in the United States of America. The text type is Meridien.
The illustrations were painted in oil on paper.

Library of Congress Cataloging in Publication Data
Johnston, Tony, 1942–
The wagon / by Tony Johnston ; paintings by James E. Ransome.—1st ed. p.   cm.
Summary: A young boy is sustained by his family as he endures the difficulties of being a slave,
but when he finally gains his freedom, his joy is tempered by the death of President Lincoln.
[1. Slavery—Fiction. 2. Afro-Americans—Fiction.] I. Ransome, James, ill. II. Title.
PZ7.J6478Wag 1996 [E]—dc20 95-53103 CIP AC
ISBN 0-688-13457-2 (trade).—ISBN 0-688-13537-4 (le)

10 9 8 7 6 5 4 3 2 1
First edition

I touch old wood.
  I touch old wounds. — T.J.

To my second blessing, Maya, with the muffin face
  and big beautiful eyes. — J.E.R.

One Carolina morning, I was born.

Everything was beautiful that day, Mama said,

especially my skin

like smooth, dark wood.

But like all my family, birth to grave,

my skin made me

a slave.

We lived on a farm, where we worked

for a man.

His house was white as ripe

cotton, the grass green

as spring.

The fields rolled on forever it seemed.

How I hated the place!

I could not go

where I pleased.

My papa was good at building things.
His hands, though huge, made magic
with tools.
Once he built a wagon, like Master
told him.
Though I had lived eight plantings only,
I helped fell a tall oak.
Then I handed Papa his needs
when he called.
"Hammer!" "Saw!" "Nail!" "Awl!"
We built a good wagon
of smooth, dark wood.

Master strolled around it, slow.
Smacked one side. Inspected the wheels
sharp and close, like an auctioneer
checks a horse—
or a slave.
I longed to climb atop that wagon and roll
away.

Master put Papa in charge of it.
Gave us the use of his mules.
In our room, cramped as a cracker box,
my family whooped
at those slave names. We pondered
a change.
At sunup, the mules had been
Robert and James. By sundown, they were
Swing and Low,
from a song we sang about a band of angels
and a better place.
I thought, *Any place is better*
than this,
where all day I heard the whish of the lash
like the Devil's breath.
When the overseer beat an old man
near to death,
I cried and got whipped,
for that.

Every day now, Papa and I were

allowed

to hitch up Swing and Low and go

for supplies, then come straight back.

No dawdling. No delays.

Or we would be lashed

to the bone.

Sometimes our cargo was slaves,

to be sold.

Sometimes, slaves coming

to the farm.

Papa sang

while the two of us creaked along.

"Swing low, sweet chariot, comin' for to

carry me home."

"What's a chariot?" I asked once.

"Something to bear you off."

"Like this wagon?"

"Like this wagon, but

glorious."

When we had set out, it was just

a wagon. Now it was my glorious chariot.

Maybe some day, it would bear me

away.

Papa was wagon boss. Also in charge of splitting
wood to warm the farm.

Though spindly as a sapling,
dawn to dark, I was to help.

How that rankled me!
I was a *boy*, not a mule.
I longed to do what free

boys do.

I envied all the footloose

things. Even the sun,

and the moon.

At times, I was surprised to feel my teeth

clench up. To feel tears warm

my face.

I brushed them away with a sleeve

and swore. *Tonight I'm going.*

Though I had the grit, I could not quit

my family. So my anger

just kept flowing.

My mama and papa knew.

"Keep courage," they said. "Trust in the Lord."

So I did what I was told.

And I grew.

Stories reached us, like smoke
on the air.
Stories of war.
In some other place, cannons growled.
Soldiers, blue and gray, prowled
other green fields.
I ached to steal away there, to fight
the Slavery Snake.
But I could not go
where I liked.

In spite of war, spring came in a burst
of blooms.
One day, I rested my ax and watched
a bug split its skin and become
a bright, flying jewel.
I wished I could split
my skin too.
I gazed at the wagon, chariot of false
hope.
In rage, I hacked
at its wheels, its spokes.

I got striped good for that.

My grandma bathed my raw back.

"Yours is not the only troubled soul," she said.

"Mr. Lincoln is sometimes overcome

with gloom.

Sees the Country ripped to rags,

as if two furious folks was tugging

at a beautiful quilt.

Sees boys dyin' and dyin'

and dead.

Does he give up?

I hear he chops wood, instead.

Like when he was young. Chopping helps him

go on."

I had a dream.
The President and I, chopping wood
together. By moonlight.
Chopping was easier after that.

As war spread, talk spread,
how the President was torn over the wrong
of slavery. How maybe he would free
all slaves.

Lord, how we prayed.
Then everything changed.
The President wrote some words one day.
We had gone to bed
slaves. But we woke up
free.

The farm came alive
with song—
one great slave voice, rejoicing
like one great hive a-hum.

I had dreamed of that moment forever,
had done my dance of jubilee
on the sly, stomping
in the wagon, spinning
like a twister, shouting
every song I knew—
a one-boy angel choir.
But when it happened, I wanted
to whisper.

My family held hands in the field
we had worked
and kneeled in the dirt and gave
thanks.

But ol' Freedom dragged her feet,
took her sweet time catching up
to those words.

When at last we left the farm, Master
was sharp-edged, but not mean.
He asked, what did we need.
Papa said, "A wagon."

So that was our farewell, the wagon
we had built long ago. And two mules,
Swing and Low.

It was eerie-quiet as we rolled toward
our new life.
We were scared. Who could see down
Freedom's furrow?
All I could hear was the snuffle
of the mules and the creak of my sweet
chariot.
I looked at my arm,
resting on the wagon flank.
It looked good—my arm
like smooth, dark wood.

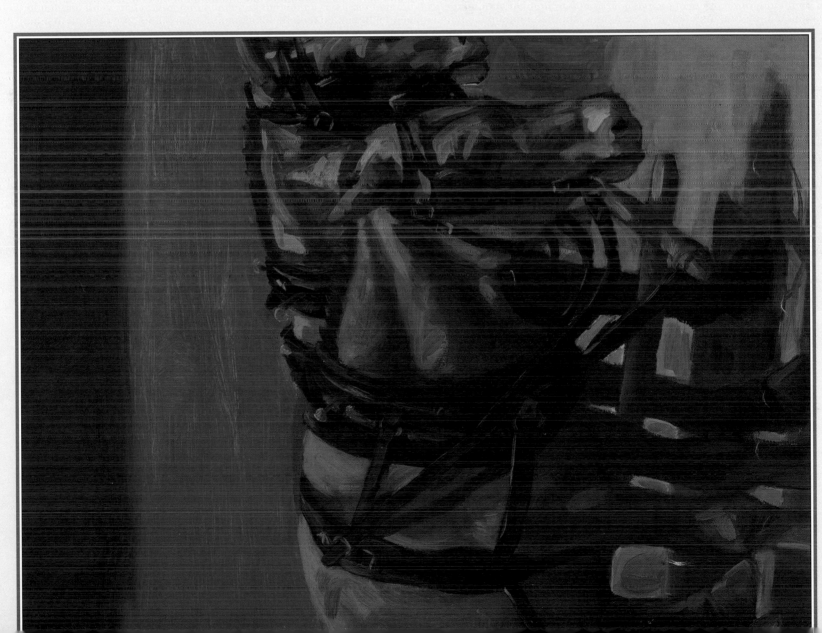

Bad news travels fast as
a snakebit wind.

One day bad news blew in.
Just when he could rejoice in war's end,
someone had shot the President
dead.

"Why?" I asked, dazed.

"Don't know," Papa said. "Seems
when you gain something of value, you must
lose something too."

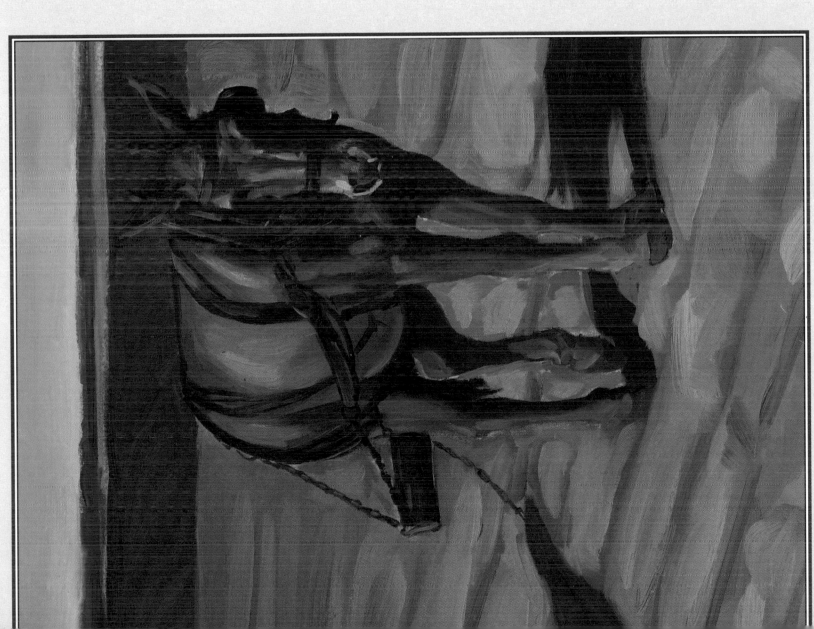

In my life, twelve plantings had come and gone.
I was free. I could go
where I pleased.
I said, "I want to go to the funeral."
So at dawn my family and I set out,
creaking down the road toward
Washington.
Creaking along in a wooden wagon,
to say goodbye to
Mr. Lincoln.

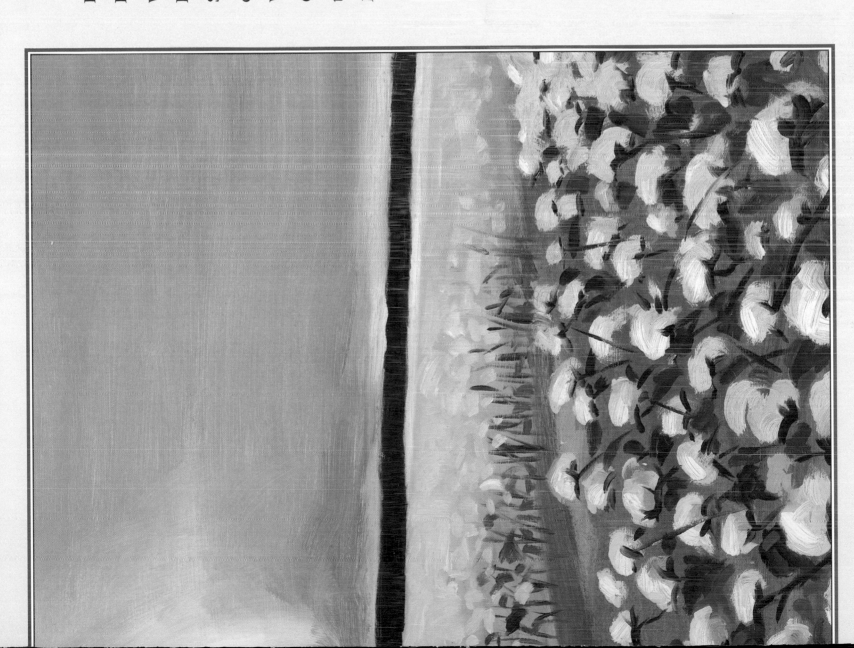